BLAKE THORNTON
Washington University in St. Louis

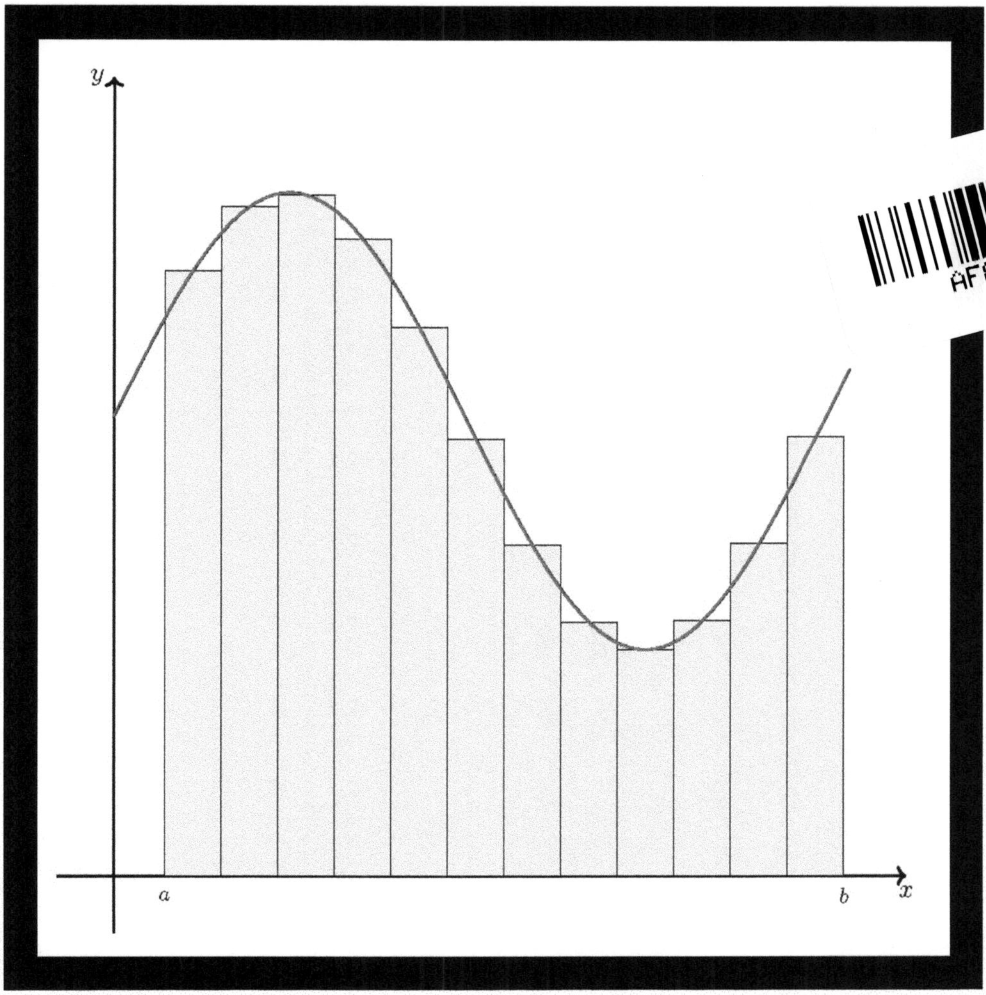

Calculus II
WORKBOOK

Kendall Hunt
publishing company

Graphs and illustrations throughout text created by Blake Thornton.

Cover image created by Blake Thornton.

Kendall Hunt
publishing company

www.kendallhunt.com
Send all inquiries to:
4050 Westmark Drive
Dubuque, IA 52004-1840

ISBN: 978-1-5249-7850-1

Published in the United States of America

Contents

Introduction

Getting students to actually do mathematics rather than watch mathematics has been a challenge for mathematics teaching for years. My approach to this problem has slowly evolved into these notes.

The basic method is simple. I want students to spend as much time in lecture actively working on problems rather than passively watching me work problems. Before class I used to print out and distribute daily problems sets that students would work through in class. These daily problems became these notes.

In lecture, I try to model a problem and then students work a similar problem. Students work together, answer each other's questions and I circulate around class helping where needed. My goal is that by the end of the class period, student leaving either having had some success or understanding where they need to invest their study time.

For Students: We will be covering these problems linearly in class. There are more problems here than we can hope to get through during class, so this gives you some extra problems to practice. I suggest you work on the problems here before your homework, to make sure you understand the main techniques we develop. There will likely be some days when you leave class confused. This realization is very useful as it should tell you to visit your instructor, teaching assistant, study group or someone who can help you get back on track.

This workbook is a work in progress. It is still rough around the edges, almost certainly with some typos. If you notice any major errors or see any corrections, please let me know and I will get those updated.

Section 1

4.9: Anti-derivatives

- Given a function, $f(x)$, find another function whose derivative is $f(x)$.
- We will often write $\int f(x)\,dx$ to mean "the most general antiderivative" of $f(x)$.

---

1. Suppose there is a function $f(x)$ and you know that $f'(x) = 6x^2 - 2x$.
   What is $f(x)$?

2. Find antiderivatives for the functions below. Be sure to double check your work by taking the derivative of your answer.

   (a) $f(x) = \sin x + x^3$

   (b) $f(x) = \frac{2}{x} - \frac{3}{x^2}$

   (c) $f(x) = 0$

   (d) $f(x) = 1$

Section 2

5.1: Areas and Distances

- Summation notation: $\sum_{i=1}^{10} a_i$
- Area under a graph of a function
- Riemann sums, left hand sums, right hand sums, mid-point sums, trapezoid sums
- $\int f(x)\,dx$ means "the most general antiderivative of $f(x)$."
  This is called an indefinite integral.
- $\int_a^b f(x)\,dx$ means "the area under the graph of $f(x)$ above the x-axis between $x = a$ and $x = b$."
  This is called a definite integral.
  (In fact, by using the definition below, the definite integral is actually the *"signed area"* under the curve–area below the x-axis counts as negative.)
- By definition, $\int_a^b f(x)\,dx$ is the limit of Riemann Sums:

$$\int_a^b f(x)\,dx = \lim_{n\to\infty} \sum_{i=1}^{n} f(x_i^*)\Delta x$$

---

1. Compute the following:

   (a) $\sum_{i=2}^{6} i^2$

   (b) $\sum_{i=1}^{5} \pi$

   (c) $\sum_{i=1}^{5} i$

   (d) $\sum_{i=1}^{5} i^2$

   (e) $\sum_{i=1}^{5} (i + i^2)$

2. Compute the area of the trapezoid:

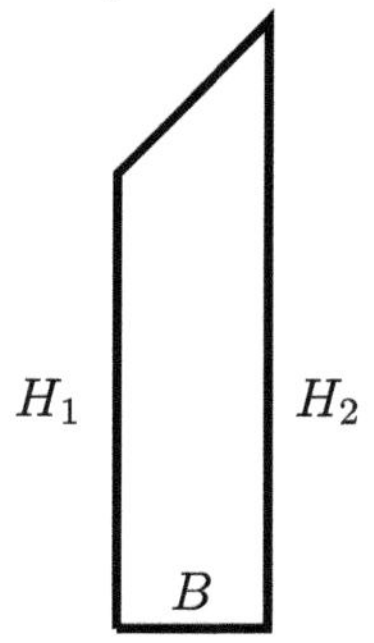

3. For $\int_2^4 (2x+3)\,dx$ compute LHS_4.

4. For $\int_2^4 (2x+3)\,dx$ compute RHS_4, T_4 and M_4.

5. If you approximate $\int_1^9 (28+x^3)\, dx$, using left hand sums, right hand sums, midpoint sums and trapezoid sums. Which of the following is the largest:

   $LHS_{32}, \quad RHS_{32}, \quad M_{32}, \quad T_{32}$

6. If you approximate $\int_1^9 (28-x^3)\, dx$, using left hand sums, right hand sums, midpoint sums and trapezoid sums. Which of the following is the largest:

   $LHS_{32}, \quad RHS_{32}, \quad M_{32}, \quad T_{32}$

7. If you approximate $\int_0^{3\pi/4} \sin x\, dx$, using left hand sums, right hand sums, midpoint sums and trapezoid sums. Which of the following is the largest:

   $LHS_{32}, \quad RHS_{32}, \quad M_{32}, \quad T_{32}$

8. In the graph below. The units of the horizontal axis, t, are seconds. The units of the vertical axis, $v(t)$, are feet per second.

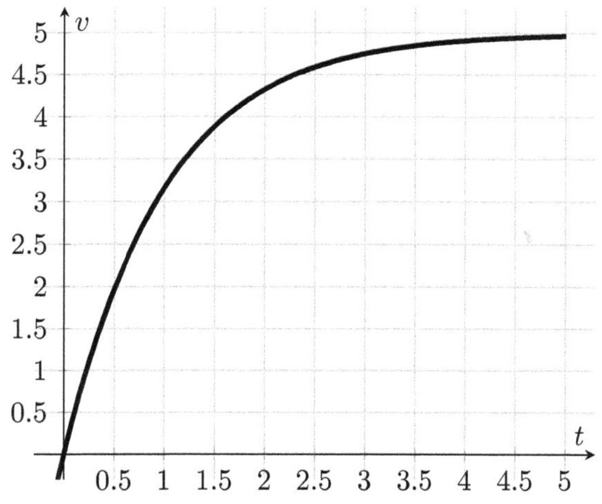

Shade in the area under the curve and above the horizontal axis between $t = 1$ and $t = 5$.

(a) What are the units of the area?

(b) Estimate the area you shaded.

9. Compute $\int_{-1}^{2} |x|\, dx$

10. $\int_{-5}^{0} \sqrt{25 - x^2}\, dx$

11. Follow the steps below to compute $\int_1^3 x^2\,dx$ using a Riemann Sum.

(a) $a =$

(b) $b =$

(c) $\Delta x =$

(d) $x_i =$

(e) $R_n =$

(f) Simplify the Riemann sum (goal: have an algebraic expression without any $\sum$ symbols)

(g) The definite integral is the limit of your Riemann sums as $n \to \infty$. Find this limit.

12. Using Riemann sums, compute $\int_0^4 x^3\,dx$

13. Using Riemann sums, compute $\int_2^5 (8x - x^2)\,dx$

Section 3

5.2: Definite Integrals

- Definition of the definite integral using Riemann Sums

$$\int_a^b f(x)\,dx = \lim_{n\to\infty} \sum_{i=1}^{n} f(x_i^*)\Delta x$$

- **Properties of the Definite Integral**

I. $\int_a^b f(x)\,dx$ is signed area.

II. $\int_a^b c\,dx = c(b-a)$

III. $\int_a^b f(x)+g(x)\,dx = \int_a^b f(x)\,dx + \int_a^b g(x)\,dx$

IV. $\int_a^b cf(x)\,dx = c\int_a^b f(x)\,dx$

V. $\int_a^b f(x);dx = -\int_b^a f(x)\,dx$

VI. $\int_a^b f(x)\,dx = \int_a^c f(x)\,dx + \int_c^b f(x)\,dx$

VII. If $f(x) \geq 0$ then $\int_a^b f(x)\,dx \geq 0$

VIII. If $f(x) \geq g(x)$ then $\int_a^b f(x)\,dx \geq \int_a^b g(x)\,dx$

IX. If $m \leq f(x) \leq M$ then $m(b-a) \leq \int_a^b f(x)\,dx \leq M(b-a)$

---

1. Suppose you know nothing about $f(x)$ except the following:
   $\int_1^3 f(x)\,dx = 2$ and $\int_3^4 f(x)\,dx = 6$.
   What is $\int_1^4 f(x)\,dx$?

2. Suppose you know nothing about $f(x)$ except the following:
   $\int_1^7 f(x)\,dx = 3$ and $\int_7^9 f(x)\,dx = -1$.
   What is $\int_1^9 f(x)\,dx$?

3. Suppose you know $\int_0^1 f(x)\,dx = 2$, $\int_1^2 f(x)\,dx = 3$, $\int_0^1 g(x)\,dx = -1$, $\int_0^2 g(x)\,dx = 4$.
   Find the following:

   (a) $\int_0^2 f(x)\,dx + \int_2^1 g(x)\,dx$

   (b) $\int_1^2 g(x)\,dx$

   (c) $\int_0^2 [2f(x) - 3g(x)]\,dx$

   (d) $\int_1^1 g(x)\,dx$

   (e) $\int_1^2 f(x)\,dx + \int_2^0 g(x)\,dx$

4. Suppose $\int_1^b f(x)\,dx = 1 - \frac{1}{b}$. Find the following:

   (a) $\int_1^5 f(x)\,dx$

   (b) $\int_{1/2}^1 f(x)\,dx$

   (c) $\int_1^6 3f(x) - 4\,dx$

   (d) $\int_3^5 f(x)\,dx$

5. The limit of sums below represents a definite integral. Find the integral. (You need to find the function and the limits of integration.)

$$\lim_{n\to\infty} \sum_{i=1}^{n} \left(e^{5+2i/n}\right) \frac{2}{n}$$

6. The limit of sums below represents a definite integral. Find the integral. (You need to find the function and the limits of integration.)

$$\lim_{n\to\infty} \sum_{i=1}^{n} (1 + 3i/n)^3 \frac{3}{n}$$

7. The limit of sums below represents a definite integral. Find the integral. (You need to find the function and the limits of integration.)

$$\lim_{n\to\infty} \sum_{i=1}^{n} (2 + 4i/n)^2 \frac{4}{5n}$$

Section 4

5.3: Fundamental Theorem of Calculus

- FTC Part 1

$$\frac{d}{dx}\int_a^x f(t)\,dt = f(x)$$

Note that we are taking a derivative of an integral.
It is also important to note here that a is a constant and that the variable, x, is the upper limit of integration.

- FTC Part 2

$$\int_a^b F'(x)\,dx = F(b) - F(a)$$

Note that we are taking an integral of a derivative.
Sometimes this is called the *Net Change Theorem.*

---

1. Chain Rule Practice. Remember that $[f(g(x))]' = f'(g(x))\,g'(x)$
   Suppose you have a function $F(x)$ and you know $F'(x) = \sin(x^2 + 1)$. Find:
   (a) $[F(x)]' =$
   (b) $\left[F(x^2)\right]' =$
   (c) $[F(\sin x)]' =$
   (d) $[F(\ln x)]' =$

2. Chain Rule Practice. Remember that $[f(g(x))]' = f'(g(x))\,g'(x)$
   Suppose you have a function $F(x)$ and you know $F'(x) = \cos(x^5 + 1)$. Find the following (hint: identify $g(x)$ in each problem and apply the chain rule).
   (a) $[F(x)]' =$
   (b) $\left[F(x^3)\right]' =$
   (c) $[F(\sin x)]' =$
   (d) $[F(\ln x)]' =$

3. Consider the function $f(t) = 2t + 1$. Let $A(x) = \int_0^x f(t)\, dt$.

(a) Find $A(3)$. What is the geometric interpretation of $A(3)$?

(b) Find $A(-3)$. What is the geometric interpretation of $A(1)$?

(c) Find a formula for $A(x)$.

(d) Find $A'(x)$.

4. Use the graph of the velocity of a car, $v(x)$, where x is in seconds and $v(x)$ is in feet per second. Let $s(x) = \int_0^x v(t)\, dt$.

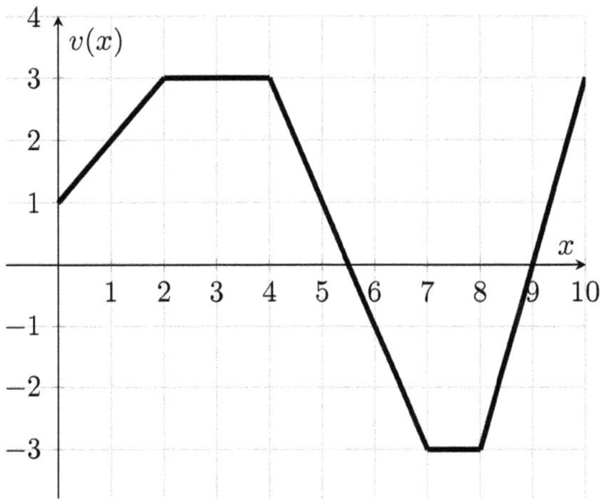

(a) Find $s(10)$.

(b) What does $s(10)$ mean physically?

(c) Let $s(x) = \int_0^x v(t)\, dt$.

(d) Draw a graph of $s(x)$.

5. Find the derivatives:

   (a) $\displaystyle \frac{d}{dx} \int_0^x \sin(t)\cos(2t)\, dt =$

   (b) $\displaystyle \frac{d}{dx} \int_2^x \sin(t)\cos(2t)\, dt =$

   (c) $\displaystyle \frac{d}{dx} \int_x^0 \sin(t)\cos(2t)\, dt =$

   (d) $\displaystyle \frac{d}{dx} \int_x^2 \sin(t)\cos(2t)\, dt =$

   (e) $\displaystyle \frac{d}{dx} \int_2^{10} \sin(x)\cos(2x)\, dx =$

6. Find the derivatives. (Hint: Use the Chain Rule.)

   (a) $\displaystyle \frac{d}{dx} \int_0^x \cos(t^5+1)\, dt =$

   (b) $\displaystyle \frac{d}{dx} \int_0^{x^3} \cos(t^5+1)\, dt =$

   (c) $\displaystyle \frac{d}{dx} \int_0^{\sin x} \cos(t^5+1)\, dt =$

   (d) $\displaystyle \frac{d}{dx} \int_0^{\ln x} \cos(t^5+1)\, dt =$

7. Find the derivatives.
   (Hint: Use properties of integrals to split up the integral.)

   (a) $\dfrac{d}{dx}\displaystyle\int_{x}^{x^2} \sin(t^2)\, dt$

   (b) $\dfrac{d}{dx}\displaystyle\int_{-x}^{3} \sin(t^2)\, dt$

   (c) $\dfrac{d}{dx}\displaystyle\int_{-x}^{\ln x} \sin(t^2)\, dt$

   (d) $\dfrac{d}{dx}\displaystyle\int_{\cos x}^{\sin x} \sin(t^2)\, dt$

8. Find functions with the given derivatives.
(Hint: use the Fundamental Theorem of Calculus, Part 2.)

(a) $f'(x) = \sin(x^2)$

(b) $f'(x) = e^{x^2}$

(c) $f'(x) = e^{\cos x}$

9. Explain the picture and the Mean Value Theorem (MVT)

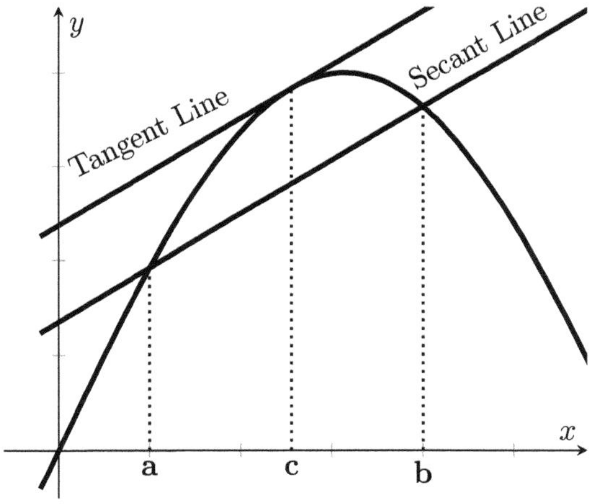

10. Find the following indefinite integrals using FTC part 2.

(a) $\int_{1}^{9} -2\, dx =$

(b) $\int_{1}^{4} x\, dx =$

(c) $\int_{-1}^{1} x^2\, dx =$

(d) $\int_{0}^{\pi} \sin x\, dx =$

(e) $\int_{1}^{10} \frac{1}{x}\, dx =$

(f) $\int_{0}^{\pi/4} \sec^2 x\, dx =$

(g) $\int_{1}^{4} x\, dx =$

11. (True/False) Any derivative formula gives a corresponding indefinite integral formula?

Section 5

5.4: Net Change Theorem (FTC Part 2)

- FTC Part 2 (also sometimes called the Net Change Theorem)

$$\int_a^b F'(x)\,dx = F(b) - F(a)$$

---

1. True or False.
   The FTC part 2 says that if you integrate velocity then you get distance travelled.

2. Let $v(t) = -\frac{3}{2}t^2 + 9t$ be velocity at time t. Find the total distance travelled between $t = 0$ and $t = 8$.

3. Suppose you are in Denver and you drive your car towards the mountains. Let x represent the number of miles from Denver and let $h(x)$ represent your height above sea level (in feet) when you are x miles from Denver. (Note: Vail is about 100 miles from Denver.)

   (a) What does $h'(x)$ represent? (Write a sentence.)

   (b) What would it mean if $h'(25) = 100$? What if $h'(30) = -12$?

   (c) What is the interpretation of $\int_0^{100} h'(x)\,dx$?

   (d) What would it mean if $\int_0^{100} h'(x)\,dx = 2690$?

   (e) What if $\int_0^{100} h'(x)\,dx = -2690$?

Section 6

5.5: Substitution

- If $u = g(x)$ then

$$\int_a^b f(g(x))g'(x)\,dx = \int_{g(a)}^{g(b)} f(u)\,du$$

- Definite integrals of even and odd functions:

$$\text{If } f(x) \text{ is odd:} \quad \int_{-a}^{a} f(x)\,dx = 0$$

$$\text{If } g(x) \text{ is even:} \quad \int_{-a}^{a} g(x)\,dx = 2\int_0^a g(x)\,dx$$

---

1. Integration practice. Compute the following indefinite integrals:

(a) $\int x^3\,dx =$

(b) $\int \frac{1}{x^3}\,dx =$

(c) $\int \frac{1}{\sqrt[3]{x}}\,dx =$

(d) $\int \frac{1}{x}\,dx =$

(e) $\int \sin x\,dx =$

(f) $\int \cos x\,dx =$

(g) $\int \sec^2 x\,dx =$

(h) $\int e^x\,dx =$

2. One goal of the class will be to use derivative formulas to find integral formulas.
The chain rule is: $[f(g(x))]' = f'(g(x))g'(x)$
Write down the corresponding integral formula.

3. Perform the given u-substitution on the given integral. Use this to convert the integral to a u-integral. If possible, compute the indefinite integral.

(a) $\int x^2 \sin(x^3 + 3)\, dx$
$u = x^3$
$du =$

(b) $\int \sin(x^3 + 3)\, dx$
$u = x^3$
$du =$

(c) $\int \tan x\, dx$
$u = \tan x$
$du =$

(d) $\int \tan x\, dx$
$u = \cos x$
$du =$

4. Compute the integrals, if possible, using the given u-substitution.

(a) $\displaystyle\int \frac{15}{3-2x}\,dx$
$u = 3 - 2x$
$du =$

(b) $\displaystyle\int \frac{1}{x^2\left(1+\frac{1}{x}\right)}\,dx$
$u = 1 + \frac{1}{x}$
$du =$

(c) $\displaystyle\int (2x+1)e^{x^2+2x+3}\,dx$
$u = x^2 + 2x + 3$
$du =$

(d) $\displaystyle\int \sqrt{7x+9}\,dx$
$u = 7x + 9$
$du =$

(e) $\displaystyle\int x\sqrt{4-x}\,dx$
$u = x$
$du =$

(f) $\displaystyle\int x\sqrt{4-x}\,dx$
$u = 4 - x$
$du =$

5. Determine the best substitution for the integral.
In order to convince yourself it is "the best" you should integrate too!

$$\int 3x^2 \csc^2(x^3+1)\, dx$$

6. Find $\displaystyle\int \frac{5\cos 5x}{e^{\sin 5x}}\, dx$ by using $u = \sin 5x$.

7. True or False. When using the u-substitution technique, sometimes it is sometimes useful to use substitution more than once.

8. Make the substitution $u = \ln x$ in the integral:

$$\int_1^e \frac{\ln x}{x}\, dx$$

9. Compute the definite integrals with substitution in two ways:

   I. Change the limits to u-limits and use these u-limits and

   II. Keep the x-limits and make sure you change back to x.

   (a) $\displaystyle\int_1^2 x\sqrt{x-1}\, dx =$

   (b) $\displaystyle\int_0^3 \frac{1}{5x+1}\, dx =$

   (c) $\displaystyle\int_1^2 \frac{e^{1/x}}{x^2}\, dx =$

   (d) $\displaystyle\int_1^3 \frac{x}{1+2x}\, dx =$

   (e) $\displaystyle\int_0^1 \frac{1}{(1+\sqrt{x})^4}\, dx =$

10. Integrate $\int_0^9 \sqrt{4 - \sqrt{x}}\, dx$

11. (a) What is an even function? Draw an example of a graph of an even function.

(b) What is an odd function? Draw an example of a graph of an odd function.

(c) Explain what happens if you want to compute the definite integral of an even or odd function from $-a$ to a for some value of a.

12. Select the best substitution for the integral.

$$\int_{-1}^{1} x^5 \sqrt[3]{x^4 + 1}\, dx$$

Section 7

6.1: Areas

- Area Between Curves
  - Graph the region
  - Find any/all intersections
  - Decide if you are going to integrate dx or dy
  - Area $= \displaystyle\int_{x=a}^{x=b} (\text{Big} - \text{Small})\, dx$

    OR

    Area $= \displaystyle\int_{y=a}^{y=b} (\text{Big} - \text{Small})\, dy$

---

1. Graph, find intersections, find the area of the region.
   $y = x^2 + x$ and $y = 4x^2 - 20x$

2. Graph, find intersections, find the area of the region.
   $y = 2x^2 + 10$ and $y = x^2 + 7x$

3. Graph, find intersections, find the area of the region.
   $x = y^2 - 4$ and $x = 2y - y^2$

4. Given the graph of the functions below, find and compute integrals that represent the area between $y^2 = 4x$ and $y = 2x - 4$. Do this both as a dx-integral and as a dy-integral.

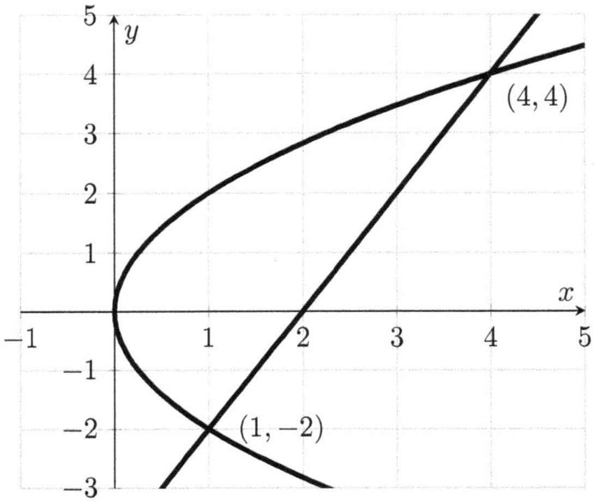

5. Find the area between $y = \sin x$ and $y = \cos x$ between $x = 0$ and $x = 2\pi$. (Be sure to graph the curves!)

6. Find the area between $y = x^2$ and $y = \sqrt{x}$.
   Do this in two ways: as a dx integral and as a dy integral.

7. Find the area between $y = xe^{-x^2}$, $y = x + 1$, $x = 2$ and the y axis.

8. Find the area between $y = 2x^2 + 10$, $y = 4x + 16$, $x = -2$ and $x = 5$.

9. Find the area enclosed by the graphs of $y = 8 - x^2$, $y = 7x$ and $y = 2x$ in the first quadrant.

Section 8

6.2: Volumes and Volumes of Revolution

- Volumes by slicing: $V = \int_a^b A(x)\,dx$ (or $A(y)\,dy$).
- Volumes of revolution using washers:

$$V = \int_a^b \pi r^2\,dx \qquad \text{OR} \qquad V = \int_a^b \pi r^2\,dy$$

1. Draw the region and draw the axis of rotation.
2. Draw the region reflected about the axis of rotation. I also like to draw a few circles showing the solid of rotation.
3. If axis is vertical, you should integrate dy.
   If axis is horizontal, you should integrate dx.
4. Identify outer and inner radii in your picture.
   Be sure you measure from the axis to the "original" region, not the reflected region.
5. Set up integral and integrate

---

1. Given two y values, y_1 and y_2, how do you find the distance between these y-values?

2. Let R be the region in quadrant I, between $y = x^2$, $x = 0$ and $y = 4$.

   (a) Graph this region.

   (b) Let S be the solid with base R and with squares as cross sections perpendicular to the x-axis. Draw S.

   (c) Find the area of the cross section at x.

   (d) Find the volume of this solid.

3. Let R be the region in the xy-plan enclosed by $x = y^2$ and $x = 3$.

   (a) Graph the region.

   (b) Let S_1 be a solid with base R. If vertical cross sections of S_1 perpendicular to the x-axis are squares, find a formula for $A(x)$, the cross sectional area at x.

   (c) Find the volume of S_1.

   (d) What do the horizontal cross sections look like in the solid S_1?

4. Let R be the region in the xy-plan enclosed by $x = y^2$ and $x = 3$.

   (a) Let S_2 be a solid with base R. If vertical cross sections of S_2 perpendicular to the x-axis are semi-circles, find a formula for $A(x)$, the cross sectional area at x.

   (b) Find the volume of S_2.

5. Let R be the region in the xy-plan enclosed by $x = y^2$ and $x = 3$.

   (a) Let S_3 be a solid with base R. If vertical cross sections of S_2 perpendicular to the x-axis are equilateral triangles, find a formula for $A(x)$, the cross sectional area at x.

   (b) Find the volume of S_3.

6. Let R be the region in quadrant I between $y = x^2$ and $y = 4$.

   (a) Graph this region.

   (b) Let S_1 be the solid of this region rotated around the y-axis. Draw S_1 and find the volume of the solid.

   (c) Let S_2 be the solid of this region rotated around the x-axis. Draw S_2 and find the volume of the solid.

   (d) Let S_3 be the solid of this region rotated around the line $x = 3$. Draw S_3 and find the volume of the solid.

   (e) Let S_4 be the solid of this region rotated around the line $y = -2$. Draw S_4 and find the volume of the solid.

7. Let R be the region between $y = x^2$ and $y = 2x$.

   (a) Graph the region R.

   (b) Let S_1 be the solid of this region rotated around the y-axis. Draw S_1 and find the volume of the solid.

   (c) Let S_2 be the solid of this region rotated around the x-axis. Draw S_2 and find the volume of the solid.

   (d) Let S_3 be the solid of this region rotated around the line $x = 3$. Draw S_3 and find the volume of the solid.

   (e) Let S_4 be the solid of this region rotated around the line $y = -2$. Draw S_4 and find the volume of the solid.

8. Let R be the region bounded by $y = x^2$ and $y = x + 2$.

   (a) Draw the region R.

   (b) Draw this region rotated about $y = -1$

   (c) Draw in a slice when $x = -1/2$.

   (d) Draw in more slices: $x = -1$, $x = 0$, $x = 1$, $x = 2$.

   (e) With your fingers on your diagram, point out the outer radius of the washer. For a given slice at x, write down a formula for the outer radius.

   (f) With your fingers on your diagram, point out the inner radius of the washer. For a given slice at x, write down a formula for the inner radius.

   (g) With your fingers, point out the x values for which you have slices.

   (h) Cross sectional area is $A(x) = \pi(r_o^2 - r_i^2)$. Find an expression for $A(x)$.

   (i) Write down an integral representing the volume and compute.

9. Take the region bounded by $y = x^2$ and $y = x + 2$. and rotate about $y = 0$ (the x-axis). Find the volume.

10. Take the region bounded by $y = x^2$ and $y = x + 2$. and rotate about $y = 4$. Find the volume.

11. Take the region bounded by $y = x^2$ and $y = x + 2$. and rotate about $y = 100$. Find the volume.

12. Take the region bounded by $y = x^2$ and $y = x + 2$. and rotate about $y = 1$. Find the volume. (Tricky!)

13. Take the region bounded by $y = x^2$ and $y = x + 2$. and rotate about $y = 3$. Find the volume. (Tricky!)

14. Let R be the region bounded by $y = x^2 + 1$ and $y = -x + 3$.

    (a) Draw the region R.

    (b) Find the volume of R rotated about the x-axis.

    (c) Find the volume of R rotated about the line $y = -5$.

    (d) Find the volume of R rotated about the line $y = -100$.

    (e) Find the volume of R rotated about the line $y = 1000$.

Section 9

6.3: Volumes by Cylindrical Shells

- Volumes of revolution (shells): $V = \int_a^b 2\pi r h \, dr$

1. Draw the region and draw the axis of rotation.
2. Draw the region reflected about the axis of rotation. I also like to draw a few circles showing the solid of rotation.
3. If axis is vertical, you should integrate dx.
   If axis is horizontal, you should integrate yx.
   (Note: this is the opposite of the washer method.)
4. Identify radius and height in your picture.
   Be sure you measure your radius from the axis to the "original" region, not the reflected region.
5. Set up integral and integrate

---

1. Let R be the region in the plane bounded by $y = x^2$ and $y = \frac{1}{2}x^2 + 2$, see below: Rotate R about the line $x = 3$ and find the volume using **washers**.

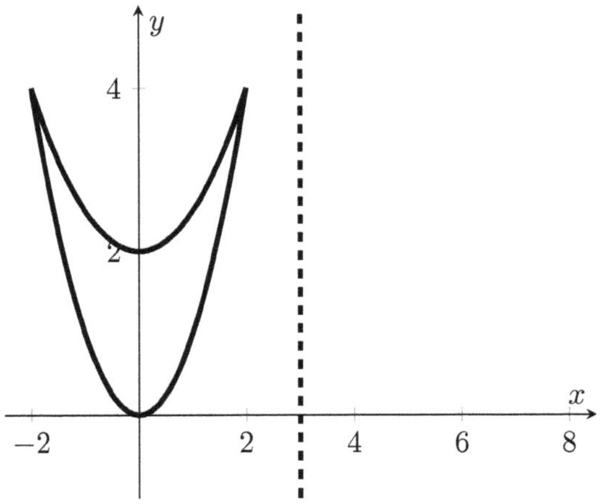

2. Find the volume of the *cylindrical shell* pictured. h is the height, r_1 is the outer radius, r_2 is the inner radii, $r = \frac{1}{2}(r_1 + r_2)$ is the average radius and $\Delta r = r_1 - r_2$.
The best answer will be in terms of r, h and Δr.

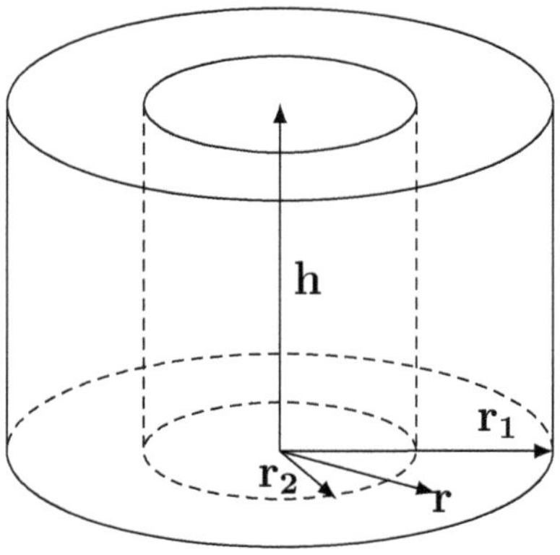

3. Let R be the region bounded by $y = 4 - x^2$ and $y = 0$.

   (a) Graph the region R.

   (b) Rotate R about the line $x = 3$ and find the volume using shells.

   (c) Find volume if R is rotated about $x = -3$.

   (d) Find volume if R is rotated about $x = 99$.

4. The graph of the region bounded by $x = y^2$ and $x = 3y^2/4 + 1$ is below.

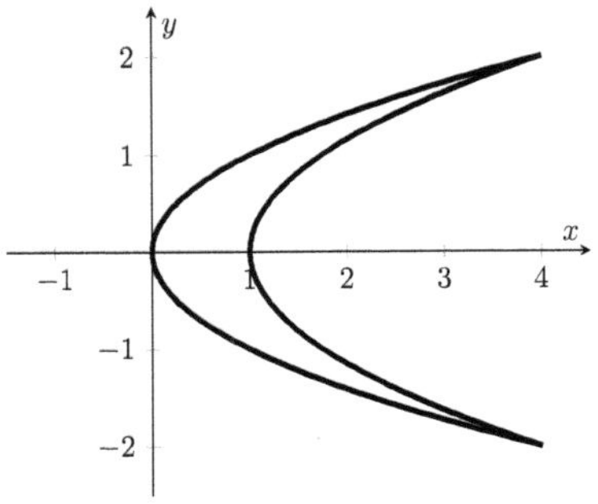

(a) Find the volume of this region rotated about $y = 3$.

(b) Find the volume if the region is rotated about the line $x = 4$.

5. Let R be the region bounded by $y = e^{-x^2}$, $y = 0$, $x = 0$ and $x = 1$. Find the volume of the solid generated by rotating R about the y-axis.

6. Let R be the region bounded by $y = \cos x$, $y = 0$, $x = 0$ and $x = \pi/2$. Let S be the solid generated by rotating R about the y-axis.

   (a) Draw the region and solid.

   (b) Using washers, write down an integral representing the volume of S.

   (c) Using shells, write down an integral representing the volume of S.

   (d) Compute the volume of S.

7. Let R be the region bounded by $y = \cos x$, $y = 0$, $x = 0$ and $x = \pi/2$. Let S be the solid generated by rotating R about the axis $x = -1$.

   (a) Draw R and S.

   (b) Set up an integral for the volume of S using washers.

   (c) Set up an integral for the volume of S using shells.

   (d) Find the volume of S.

8. Set up the integral for volume (using both shells and washers) for the solid obtained by rotating the region bounded by $x = (y-2)^2$ and $y = x$ about the line $y = -1$.

Section 10

6.5: Average Value and MVT for Integrals

- Average Value

$$f_{\text{ave}} = \frac{1}{b-a}\int_a^b f(x)\,dx$$

- Mean Value Theorem (MVT) for integrals
  If f is continuous on $[a, b]$ then there is a number $c \in [a, b]$ such that:

$$f(c) = f_{\text{ave}} = \frac{1}{b-a}\int_a^b f(x)\,dx$$

$$\int_a^b f(x)\,dx = f(c)(b-a)$$

---

1. Let $f(x) = x^2$.

   (a) For $i = 1, 2, \ldots, 5$:

i	1	2	3	4	5
$\mathbf{x_i}$	1	2	3	4	5
$\mathbf{f(x_i)}$	1^2	2^2	3^2	4^2	5^2

   Find the average of the values $f(x_1)$, $f(x_2)$, $f(x_3)$, $f(x_4)$ and $f(x_5)$ in the chart.

   (b) Now we subdivide further. For $i = 1, 2, \ldots, 10$:

i	1	2	3	4	5	6	7	8	9	10
$\mathbf{x_i}$	.5	1	1.5	2	2.5	3	3.5	4	4.5	5
$\mathbf{f(x_i)}$	$.5^2$	1^2	1.5^2	2^2	2.5^2	3^2	3.5^2	4^2	4.5^2	5^2

   Find the average of the values $f(x_i)$ in the chart.

   (c) Now we subdivide further. For $i = 1, 2, \ldots, 50$:

i	1	2	3	⋯	⋯	49	50
$\mathbf{x_i}$	0.1	0.2	0.2	⋯	⋯	4.9	5.0
$\mathbf{f(x_i)}$	$(0.1)^2$	$(0.2)^2$	$(0.3)^2$	⋯	⋯	$(4.9)^2$	$(5.0)^2$

   Find the average of the values $f(x_i)$ in the chart.

2. Find the average value.

   (a) x^2 on $[-1, 1]$

   (b) e^x on $[1, 10]$

   (c) $\sin x$ on $[0, \pi]$

   (d) $\cos x$ on $[0, \pi]$

   (e) $\cos x$ on $[0, \pi/2]$

3. For the functions and intervales in Problem 2, find the x-value, $x = c$, guaranteed by the MVT, such that $f(c) = f_{\text{ave}}$.

4. Suppose $f(x)$ is continuous on $[-1, 1]$ and $\int_{-1}^{1} f(x)\, dx = \pi$. Which, if any, of the following values does f have to take on for x in the interval $[-1, 1]$?
(Which of the following are guaranteed to be y values?)

$$1, \quad -1, \quad \pi, \quad \frac{\pi}{2}$$

5. Let $f(x) = x^2 - 8$.

   (a) Find the average value of f on $[-3, 1]$.

   (b) Find c in $[-1, 3]$ such that $f(c)$ equals the average value. (This is the value guaranteed by the MVT for integrals.)

   (c) Draw a graph of $f(x)$ with the average value labeled, the point c labeled and draw an appropriate rectangle.

6. Find the average value of $f(x) = |x^2 - 2|$ on $[0, 2]$.

Section 11

7.1: Integration by Parts

- Integration by Parts:

$$\int u\,dv = uv - \int v\,du$$

---

1. If u and v are functions and "d" means derivative, the differential form of the product rule can be written as

$$d(uv) = v\,du + u\,dv$$

Write this equation in integral form.

2. Use integration by parts to compute the integrals. Use the given u and dv.

(a) $\int \ln x\,dx$
$u = \ln x,\ dv = dx$

(b) $\int x \cos x\,dx$
$u = x,\ dv = \cos x\,dx$

(c) $\int x \cos x\,dx$
$u = \cos x,\ dv = x\,dx$

(d) $\int x \cos x\,dx$
$u = \cos x,\ dv = \quad dx$

3. Use integration by parts to compute the definite integral.

$$\int_1^4 xe^x \, dx$$

4. Use integration by parts to find the indefinite integrals.

(a) $\int x^2 e^{3x} \, dx$

(b) $\int x^2 \sin(10x) \, dx$

(c) $\int (\ln x)^2 \, dx$

(d) $\int x^3 \cos(x^2) \, dx$

5. Use integration by parts to find the indefinite integral. Use $u = e^x$ and $dv = \sin x \, dx$.

$$\int e^x \sin x \, dx$$

6. More integration by parts practice.

   (a) $\int e^x \cos x \, dx$

   (b) $\int e^{3x} \cos 4x \, dx$

7. More integration by parts practice.

(a) $\int x^2 e^{3x}\,dx$

(b) $\int x^3 e^{3x}\,dx$

(c) $\int \arctan x\,dx$

(d) $\int x \sin x \cos x\,dx$

(e) $\int x \ln x\,dx$

(f) $\int \frac{\ln x}{x^5}\,dx.$

(g) $\int 2x \arctan x\,dx$

Section 12

7.2: Trigonometric Integrals

- Trigonometric Identities
  - $\sin^2\theta + \cos^2\theta = 1$
  - $1 + \tan^2\theta = \sec^2\theta$
  - $1 + \cot^2\theta = \csc^2\theta$
  - $\sin(A \pm B) = \sin A \cos B \pm \sin B \cos A$
  - $\cos(A \pm B) = \cos A \cos B \mp \sin A \sin B$
  - $\tan(A \pm B) = \frac{\tan A \pm \tan B}{1 \mp \tan A \tan B}$
  - $\sin A \sin B = \frac{1}{2}[\cos(A - B) - \cos(A + B)]$
  - $\cos A \cos B = \frac{1}{2}[\cos(A - B) + \cos(A + B)]$
  - $\sin A \cos B = \frac{1}{2}[\sin(A + B) + \sin(A - B)]$
  - $\sin^2 x = \frac{1}{2}(1 - \cos 2x)$
  - $\cos^2 x = \frac{1}{2}(1 + \cos 2x)$
  - $\sin(2\theta) = 2\sin\theta\cos\theta$
  - $\cos(2\theta) = \cos^2\theta - \sin^2\theta$
- Trigonometric Integral Stategies
  - $\int \sin^m x \ \cos^n x \ dx$
  - $\int \tan^m x \ \sec^n x \ dx$
  - $\int \tan x \ dx$
  - $\int \sin(nx) \ \cos(mx) \ dx$
  - $\int \sec x \ dx$
  - $\int \sec^3 x \ dx$
  - Other ad-hoc methods.

---

1. Find the derivatives.
   (These will be important for our formulas for $\int \sec x \ dx$ and $\int \csc x \ dx$.)

   (a) $\frac{d}{dx}(\sec x + \tan x)$

   (b) $\frac{d}{dx}(\csc x + \cot x)$

2. Compute $\int \sec x \, dx$.
   (Try $u = \sec x + \tan x$)

3. Compute $\int \csc x \, dx$
   (Try $u = \csc x + \cot x$)

4. Use integration by parts to compute:

$$\int \sec^3 x \, dx$$

5. Integrate

$$\int \sin x \, \cos x \, dx$$

6. Integrate

$$\int \sin^4 x \, \cos^5 x \, dx$$

7. Integrate

$$\int \sin^5 x \, \cos^4 x \, dx$$

8. Integrate $\int \sin^4 x \ \cos^6 x \ dx$

9. Integrate $\int \sin^3 x \cos^5 x \ dx$

10. Integrate $\int \cos^2 x \ dx$

11. Integrate $\int \sin^3 x \cos^2 x \, dx$

12. Integrate $\int \sin^2 x \cos^7 x \, dx$

13. Integrate $\int \sin^4 x \, dx$

14. Integrate $\int \sin^6 x \, \cos^2 x \, dx$

15. Use $u = \tan x$ to integrate $\int \tan x \, \sec^6 x \, dx$

16. Use $u = \sec x$ to integrate $\int \tan x \, \sec^6 x \, dx$

17. Integrate $\int \sec^4 x \, dx$

18. Integrate $\int \tan^4 x \, dx$

19. Integrate $\int \tan^2 x \, dx$

20. Integrate $\int \sin(2x)\sin(17x)\,dx$

21. Integrate $\int \sin(5x)\cos(13x)\,dx$

22. Integrate $\int \cos(32x)\cos(3x)\,dx$

23. Integrate $\int \frac{\cos^2 x}{1 + \sin x}\, dx$

24. Integrate $\int (\sec^2 x)\sqrt{5 + \tan x}\, dx$

25. Integrate $\int \frac{\cos^5 x \ \sin x}{1 - \sin^2 x}\, dx$

26. Integrate $\int \frac{\sin 2x}{\cos^4 x + \sin^4 x}\, dx$

Section 13

7.3: Trigonometric Substitution

- Steps in Trigonometric Substitution:

  1. If necessary, complete the square.
  2. Determine the correct trig substitution.

Trigonometric Subsitution		
Expression	**Substitution**	**Identity**
$\sqrt{a^2 - x^2}$	$x = a \sin\theta$	$1 - \sin^2\theta = \cos^2\theta$
$\sqrt{a^2 + x^2}$	$x = a \tan\theta$	$1 + \tan^2\theta = \sec^2\theta$
$\sqrt{x^2 - a^2}$	$x = a \sec\theta$	$\sec^2\theta - 1 = \tan^2\theta$

  3. Substitute (don't forget dx, $d\theta$)
  4. Simplify and integrate.
  5. Definite Integral: change limits.
  6. Indefinite Integral: Change answer from θ back to x.
     Draw the trig subsitution triangle to do this.

---

1. Given the triangle below, find the values of the trig functions asked for.

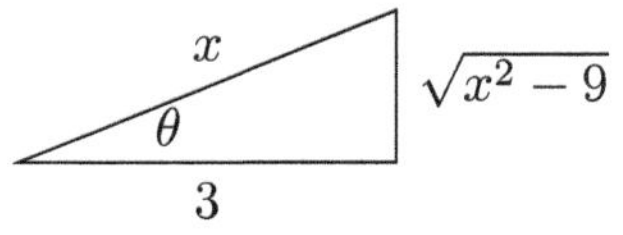

   (a) $\sin\theta$

   (b) $\cos\theta$

   (c) $\tan\theta$

   (d) $\csc\theta$

   (e) $\sec\theta$

   (f) $\cot\theta$

2. In the expression $x\sqrt{1-x^2}$, choose a substitution below that will eliminate the square root. Simplify the result.

   Possible Substitutions: $x = \sin\theta$, $x = \cos\theta$, $x = \tan\theta$, $x = \sec\theta$

3. In the expression $\dfrac{x^2}{\sqrt{x^2-1}}$, choose a substitution below that will eliminate the square root. Simplify the result.

   Possible Substitutions: $x = \sin\theta$, $x = \cos\theta$, $x = \tan\theta$, $x = \sec\theta$

4. In the expression $\dfrac{\sqrt{1+x^2}}{x^2}$, choose a substitution below that will eliminate the square root. Simplify the result.

   Possible Substitutions: $x = \sin\theta$, $x = \cos\theta$, $x = \tan\theta$, $x = \sec\theta$

5. Integrate $\displaystyle\int \frac{\sqrt{9-4x^2}}{x}\,dx$

(a) Using the chart, determine the best substitution.

(b) Draw the triangle associated to your substitution.

(c) Using your triangle, find the following (these are just practice to use your triangle):

i. $\sin\theta$

ii. $\cos\theta$

iii. $\tan\theta$

iv. $\csc\theta$

v. $\sec\theta$

vi. $\cot\theta$

(d) Use your substitution and compute the integral.

6. More trig substitution practice:

(a) $\displaystyle\int \frac{dx}{(x^2+9)^{3/2}}$

(b) $\displaystyle\int \frac{\sqrt{25x^2-4}}{x}\,dx$

(c) $\displaystyle\int_{2/5}^{4/5} \frac{\sqrt{25x^2-4}}{x}\,dx$

7. Integrate $\displaystyle\int_3^5 \frac{1}{\sqrt{x^2 - 6x + 13}}\,dx$

   (a) Complete the square $x^2 - 6x + 13 =$

   (b) Use the completed square and integrate.

8. Integrate $\displaystyle\int \sqrt{x^2 + 4x + 5}\,dx$

9. Integrate $\int_0^1 \sqrt{4 - x^2}\, dx$

10. Integrate $\int_1^2 \frac{1}{x^2\sqrt{x^2 + 16}}\, dx$

11. Integrate $\int_0^4 x^2\sqrt{16 - x^2}\, dx$

Section 14

7.4: Partial Fractions

- Partial Fractions: An algebraic technique that helps when integrating rational functions.
- Steps in Partial Fractions
  1. Long Division if deg(Numerator) $\geq$ deg(Denominator).
  2. Factor the denominator in to linear and quadratic irreducibles factors.
  3. Determine the "form" of the partial fraction decomposition.
  4. Solve for constants in your form.
  5. Integrate term by term

---

1. What is a rational function?

2. Solve the system of equations

$$\begin{aligned} 2A + 2B + C &= 13 \\ 3A - B + 2C &= -6 \\ -2A &= 1 \end{aligned}$$

3. Perform polynomial long division and write your answers correctly.

   (a) $\dfrac{x^2+1}{x+2}$

   (b) $\dfrac{x^2+1}{x^2+2}$

   (c) $\dfrac{2x^3 - x^2 + 5x + 1}{3x+2}$

   (d) $\dfrac{2x^5 - x^2 + 5x + 1}{x^3 - 2x^2 + x - 2}$

4. Find the integrals below where a, b are constants.
These are the types of integrals that will result from partial fractions.

(a) $\displaystyle\int \frac{1}{ax+b}\,dx$

(b) $\displaystyle\int \frac{1}{(ax+b)^2}\,dx$

(c) $\displaystyle\int \frac{x}{x^2+a^2}\,dx$

(d) $\displaystyle\int \frac{x}{(x^2+a^2)^2}\,dx$

(e) $\displaystyle\int \frac{1}{x^2+a^2}\,dx$

(f) $\displaystyle\int \frac{1}{(x^2+a^2)^2}\,dx$

5. Integrate with partial fractions. $\int \frac{1}{(x-1)(x-2)}\,dx$

6. Integrate with partial fractions. $\int \frac{1}{(x-1)(x-2)^2}\,dx$

7. Integrate with partial fractions. $\int \frac{1}{(x-1)(x^2+4)}\,dx$

8. Integrate with partial fractions. $\int \frac{2}{(x-1)(x^2+4)^2}\,dx$

9. Integrate using partial fractions.

(a) $\displaystyle\int \frac{x+3}{x(3x+2)}\,dx$

(b) $\displaystyle\int \frac{x+3}{x(3x+2)^3}\,dx$

(c) $\displaystyle\int \frac{x+3}{(3x+2)^3}\,dx$

(d) $\displaystyle\int \frac{3}{(3x+2)^3}\,dx$

(e) $\displaystyle\int \frac{x^2+x+1}{x(x^2+9)}\,dx$

(f) $\displaystyle\int \frac{x^2+x+1}{x(x^2+9)(x^2+4)}\,dx$

10. Solve the partial fraction decomposition

$$\frac{x^2+x+2}{(x-2)(x-1)(x+1)} = \frac{A}{x-2} + \frac{B}{x-1} + \frac{C}{x+1}$$

11. Integrate $\displaystyle\int \frac{x^5-5x^3+x^2+5x+2}{x^3-2x^2-x+2}\,dx$
(This is related to Question 10.)

Note: Problems 12-17 are linked (the corresponding parts correspond).

12. Long division practice. Perform long division and write your answer appropriately.

(a) $\dfrac{6x^4+3x^3-172x^2+102x-329}{x^2+x-30}$

(b) $\dfrac{3x^5-2x^4+13x^3-8x^2+5x+1}{x^3+4x}$

(c) $\dfrac{x^8+4x^6-x^5-4x^3+x+1}{x^5+4x^3}$

(d) $\dfrac{2x^5-4x^3+3x^2+3x+1}{x^4-2x+1}$

(e) $\dfrac{2x^6-2x^5+20x^4-20x^3+21x^2-17x+1}{x^6-x^5+10x^4-10x^3+9x^2-9x}$

13. Factoring Practice. Factor and write your answer appropriately.

(a) $x^2 + x - 30$

(b) $x^3 + 4x$

(c) $x^5 + 4x^3$

(d) $x^4 - 2x^2 + 1$

(e) $x^6 - x^5 + 10x^4 - 10x^3 + 9x^2 - 9x$

14. Write down the partial fraction form.

(a) $\dfrac{x+1}{(x-5)(x+6)}$

(b) $\dfrac{x+1}{x(x^2+4)}$

(c) $\dfrac{x+1}{x^3(x^2+4)}$

(d) $\dfrac{3x^2+x+1}{(x-1)^2(x+1)^2}$

(e) $\dfrac{3x^2+x+1}{x(x-1)(x^2+9)(x^2+1)}$

15. Find the system of equations to solve.

(a) $\dfrac{x+1}{(x-5)(x+6)} = \dfrac{A}{x-5} + \dfrac{B}{x+6}$

(b) $\dfrac{x+1}{x(x^2+4)} = \dfrac{A}{x} + \dfrac{Bx+C}{x^2+4}$

(c) $\dfrac{x+1}{x^3(x^2+4)} = \dfrac{A}{x} + \dfrac{B}{x^2} + \dfrac{C}{x^3} + \dfrac{Dx+E}{x^2+4}$

(d) $\dfrac{3x^2+x+1}{(x-1)^2(x+1)^2} = \dfrac{A}{x-1} + \dfrac{B}{(x-1)^2} + \dfrac{C}{x+1} + \dfrac{D}{(x+1)^2}$

(e) $\dfrac{3x^2+x+1}{x(x-1)(x^2+9)(x^2+1)} = \dfrac{A}{x} + \dfrac{B}{x-1} + \dfrac{Cx+D}{x^2+1} + \dfrac{Ex+F}{x^2+9}$

16. Solve the system of equations

(a) $6A - 5B = 1$
$A + B = 1$

(b) $A + B = 0$
$C = 1$
$4A = 1$

(c) $A + D = 0$
$B + E = 0$
$4A + C = 0$
$4B = 1$
$4C = 1$

(d) $A + C = 0$
$A + B - C + D = 3$
$-A + 2B - C - 2D = 1$
$-A + B + C + D = 1$

(e) $A + B + C + E = 0$
$-A - C + D - E + F = 0$
$10A + 10B + 9C - D + E - F = 0$
$-10A - 9C + 9D - E + F = 3$
$9A + 9B - 9D - F = 1$
$-9A = 1$

17. Integrate.

(a) $\displaystyle\int \left(6x^2 - 3x + 11 + \frac{6}{11(x-5)} + \frac{5}{11(x+6)}\right)\,dx$

(b) $\displaystyle\int \left(3x^2 - 2x + 1 + \frac{1}{4x} - \frac{x}{4(x^2+4)} + \frac{1}{x^2+4}\right)\,dx$

(c) $\displaystyle\int \left(x^3 - 1 - \frac{1}{16x} + \frac{1}{4x^2} + \frac{1}{4x^3} + \frac{x}{16(x^2+4)} - \frac{1}{4(x^2+4)}\right)\,dx$

(d) $\displaystyle\int \left(2x + \frac{1}{2(x-1)} + \frac{5}{4(x-1)^2} - \frac{1}{2(x+1)} + \frac{3}{4(x+1)^2}\right)\,dx$

(e) $\displaystyle\int \left(2 - \frac{1}{9x} + \frac{1}{4(x-1)} - \frac{3x}{16(x^2+1)} + \frac{1}{16(x^2+1)} + \frac{7x}{144(x^2+9)} - \frac{5}{16(x^2+9)}\right)\,dx$

Section 15

7.5: Integration Strategies

- Integration Strategies
  1. Substitution (Chain Rule)
  2. Parts (Product Rule)
  3. Trig Integrals
  4. Trig Sub
  5. Partial Fractions
  6. Tables and CAS
  7. Numerical Methods

Integrate.

1. $\displaystyle\int e^{x+e^x}\,dx$

2. $\displaystyle\int t\sec t\tan t\,dt$

3. $\displaystyle\int \frac{2x-3}{x^3+3x}\,dx$

4. $\displaystyle\int \frac{\ln(\tan x)}{\sin x\cos x}\,dx$

5. $\displaystyle\int \frac{1+\tan x}{\sec x - 1}\,dx$

6. $\displaystyle\int \sqrt{3-2x-x^2}\,dx$

7. $\displaystyle\int \frac{x^2}{x^2+1}\,dx$

8. $\displaystyle\int \frac{\cos(\frac{1}{x})}{x^3}\,dx$

9. $\displaystyle\int \frac{\sin^3 x}{\cos x}\,dx$

10. $\displaystyle\int \frac{\tan^{-1}(x)}{x^2}\,dx$

Section 16

4.4: L'Hopital's Rule and Indeterminate Forms

- L'Hopital's Rule
  - $\left(\frac{0}{0}\right)$: If $\lim_{x\to\infty} f(x) = \lim_{x\to\infty} g(x) = 0$ then
    $\lim_{x\to\infty} \frac{f(x)}{g(x)} = \lim_{x\to\infty} \frac{f'(x)}{g'(x)}$
  - $\left(\frac{\infty}{\infty}\right)$: If $\lim_{x\to\infty} f(x) = \lim_{x\to\infty} g(x) = \infty$ then
    $\lim_{x\to\infty} \frac{f(x)}{g(x)} = \lim_{x\to\infty} \frac{f'(x)}{g'(x)}$
- List of Indeterminate Forms: $\frac{0}{0}$, $\frac{\infty}{\infty}$, $0 \cdot \infty$, $\infty - \infty$, 0^0, 1^∞, ∞^0.
- List of Indeterminate Forms handled by L'Hopital: $\frac{0}{0}$, $\frac{\infty}{\infty}$

---

1. Use L'Hopital's Rule, if necessary. Make sure you identify the form of the limit!

   (a) $\lim_{x\to 0} \frac{e^x - 1}{2x}$

   (b) $\lim_{x\to \infty} \frac{e^x - 1}{2x}$

   (c) $\lim_{x\to 0} \frac{e^x}{2x}$

   (d) $\lim_{x\to \infty} \frac{e^x}{2x}$

2. Which limits below are indeterminant forms? Compute the limits using L'Hopital's Rule.

   (a) $\lim_{x \to 4} \frac{\sin x}{x}$

   (b) $\lim_{x \to 0} \frac{\sin x}{x}$

   (c) $\lim_{x \to 0} \frac{\cos x}{x}$

   (d) $\lim_{x \to \infty} xe^{-x}$

3. Compute $\lim_{x \to \infty} \frac{\ln x}{x}$

4. Compute $\lim_{x \to \infty} \frac{(\ln x)^2}{x}$

Section 17

7.8: Improper Integrals

- Improper Integrals
  - Type I: One limit of integration is ∞.
  - Type II: Vertical Asymptote
  - Use a limit to resolve the improper integral.
  - Use a separate limit for each improper piece if there are multiple improper pieces.

---

1. We are going to compute $\int_1^\infty \frac{1}{x^2}\, dx$.

   (a) Fix a number $t \geq 1$. Compute the integral $\int_1^t \frac{1}{x^2}\, dx$.

   (b) Using your previous answer, compute: $\lim_{t\to\infty} \int_1^t \frac{1}{x^2}\, dx$.

2. Find the following, if they converge

(a) $\int_1^\infty \frac{1}{x^3}\, dx$

(b) $\int_1^\infty \frac{1}{x^2}\, dx$

(c) $\int_1^\infty \frac{1}{x^{3/2}}\, dx$

(d) $\int_1^\infty \frac{1}{x^1}\, dx$

(e) $\int_1^\infty \frac{1}{x^{3/4}}\, dx$

(f) $\int_1^\infty \frac{1}{x^{1/2}}\, dx$

(g) $\int_1^\infty \frac{1}{x^0}\, dx$

(h) $\int_1^\infty x\, dx$

3. For which values of p, does $\int_1^\infty \frac{1}{x^p}\,dx$ converge?
   You should first "guess" an answer, using the examples from the previous problem. Then, once you have a guess, try to develop an argument as to why your guess is correct (also called a *proof*).

4. Find the following, if they converge.
   (a) $\int_0^1 \frac{1}{x^3}\,dx$

   (b) $\int_0^1 \frac{1}{x^2}\,dx$

   (c) $\int_0^1 \frac{1}{x^{3/2}}\,dx$

   (d) $\int_0^1 \frac{1}{x^1}\,dx$

   (e) $\int_0^1 \frac{1}{x^{3/4}}\,dx$

   (f) $\int_0^1 \frac{1}{x^{1/2}}\,dx$

   (g) $\int_0^1 \frac{1}{x^0}\,dx$

   (h) $\int_0^1 x\,dx$

5. For which values of p, does $\int_0^1 \frac{1}{x^p}\,dx$ converge?
You should first "guess" an answer, using the examples from the previous problem. Then, once you have a guess, try to develop an argument as to why your guess is correct (also called a *proof*).

6. Determine if the following improper integrals converge. If they converge, determine their value.

(a) $\int_1^\infty xe^{-x}\,dx$

(b) $\int_0^\infty \sin x\,dx$

7. Determine if the improper integrals converge.

(a) $\int_1^\infty \frac{1}{x^3}\,dx$

(b) $\int_1^\infty \frac{1}{xe^x}\,dx$

(c) $\int_1^\infty \frac{1}{x}\,dx$

(d) $\int_2^\infty \frac{1}{\ln x}\,dx$

(e) $\int_2^\infty \frac{1}{(\ln x)^2}$

8. Below is a graph of $f(x) = \frac{x}{9+x^2}$.

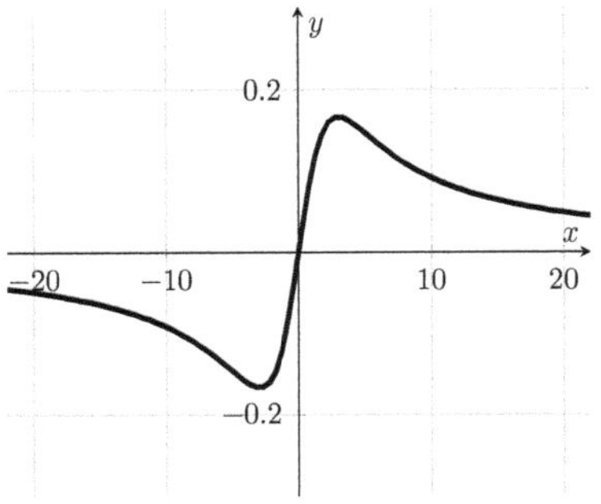

Consider the following work:

$$\int_0^{\infty} \frac{x}{9+x^2}\,dx = \lim_{t\to\infty} \int_0^t \frac{x}{9+x^2}\,dx = \lim_{t\to\infty} \frac{1}{2}\ln(x^2+9)\Big|_0^t$$
$$= \lim_{t\to\infty} \frac{1}{2}\ln(t^2+9) - \frac{1}{2}\ln(9) = \lim_{t\to\infty} \frac{1}{2}\left[\ln\left(\frac{t^2+9}{9}\right)\right] = \infty$$

$$\lim_{t\to\infty} \int_{-t}^{t} \frac{x}{9+x^2}\,dx = \lim_{t\to\infty} \frac{1}{2}\ln(x^2+9)\Big|_{-t}^{t} = \lim_{t\to\infty} \left(\frac{1}{2}\ln(t^2+9) - \frac{1}{2}\ln(t^2+9)\right) = 0$$

Does the integral converge or diverge: $\displaystyle\int_{-\infty}^{\infty} \frac{x}{9+x^2}\,dx$

9. Does the integral converge or diverge: $\displaystyle\int_{-\infty}^{\infty} \frac{1}{9+x^2}\,dx$

10. Determine if the following improper integrals converge. If they converge, determine their value.

(a) $\displaystyle\int_0^3 \frac{1}{\sqrt{3-x}}\,dx$

(b) $\displaystyle\int_0^3 \frac{1}{\sqrt{9-x^2}}\,dx$

(c) $\displaystyle\int_0^1 \ln x\,dx$

11. For the following improper integrals, determine an easier function to compare the integrand to, make the comparison, then determine convergence or divergence.

(a) $\displaystyle\int_2^\infty \frac{\cos^2 x}{x^2}\, dx$

(b) $\displaystyle\int_3^\infty \frac{1}{x + e^x}\, dx$

(c) $\displaystyle\int_2^\infty \frac{1}{\ln x}\, dx$

(d) $\displaystyle\int_2^\infty \frac{1}{\ln x - 1}\, dx$

(e) $\displaystyle\int_5^\infty \frac{x^2 + x + 1}{x^4 + x^2 + 3}\, dx$

(f) $\displaystyle\int_5^\infty \frac{x^2 + x + 1}{x^3 + x^2 + 3}\, dx$

12. Determine convergence or divergence

(a) $\displaystyle\int_5^8 \frac{6}{\sqrt{t-5}}\,dt$

(b) $\displaystyle\int_0^1 \frac{dx}{\sqrt{x^3+x}}$

(c) $\displaystyle\int_0^1 \frac{dx}{\sqrt{2x-x^2}}$

(d) $\displaystyle\int_0^\infty \frac{\sin 3x}{x^2+1}\,dx$

(e) $\displaystyle\int_1^\infty \frac{2}{x^{11}+x^{10}}\,dx$

(f) $\displaystyle\int_1^\infty \frac{1}{\sqrt{x^2+x}}\,dx$

Section 18

8.1: Arc Length

- Arc Length

$$L = \int_a^b \sqrt{1 + (f'(x))^2}\, dx = \int_a^b \sqrt{1 + \left(\frac{dy}{dx}\right)^2}\, dx = \int_a^b \sqrt{1 + \left(\frac{dx}{dy}\right)^2}\, dy$$

---

1. Find the length of the line segment between the points $(x_i, f(x_i))$ and $(x_j, f(x_j))$.
   Ideally, write your answer in terms of Δx and Δy where $\Delta x = x_j - x_i$ and $\Delta y = f(x_j) - f(x_i)$.

2. Set up and solve the arc length:

   (a) Find length of $y = 2x + 3$ between $x = 0$ and $x = 2$.

   (b) Find length of $y = \cos x$ between $x = 0$ and $x = \pi$.

   (c) Find length of $y = 2(x-1)^{3/2}$ from $x = 1$ to $x = 5$.

3. Compute arc length

   (a) Length of $y = x^3$ from $x = 1$ to $x = 4$.

   (b) Length of $x = y^3$ from $y = 1$ to $y = 4$.

   (c) Length of $y = \frac{1}{6}x^3 + \frac{1}{2x}$ from $x = 1$ to $x = 2$

4. Compute the arc length integral of $y = \ln\cos(x)$ on $[0, \pi/4]$.

Section 19

8.2: Surface Area

- Surface Area

$$S = \int_a^b dS = \int_a^b 2\pi r \, ds = \int_a^b 2\pi r \sqrt{1 + \left(\frac{dy}{dx}\right)^2} \, dx$$

- Generally we consider "Surface Area" and "Area" to be synonyms.

---

1. Find the surface area of the cone as a function of l and r. (Just the cone part, not the bottom.) Hint: Cut along the dashed line.

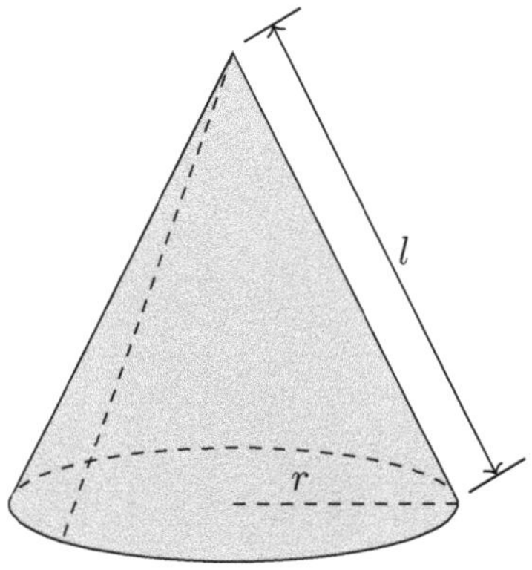

2. Find the surface area of the frustum of the cone. (Just the hollow cone part, not the top/bottom.)

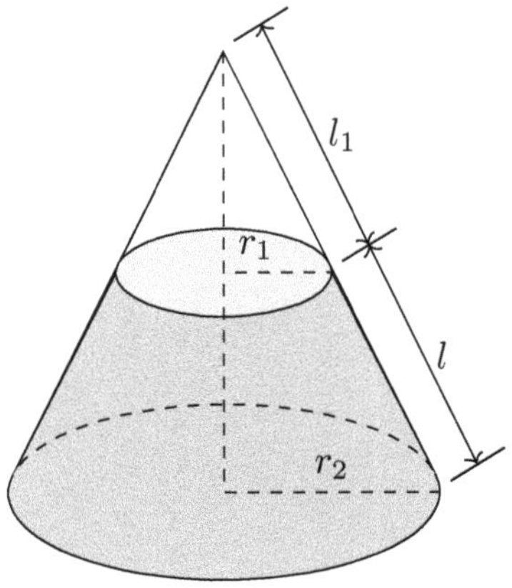

3. Find the surface area.

   (a) Find the area of $y = \sqrt{x}$ on the interval $[3/4, 15/4]$ rotated about the x-axis.

   (b) Find the area of $x = \frac{y^3}{3}$ on the interval $y = 0$ to $y = 1$ rotated about the y-axis.

   (c) Find the area of $y = \sin x$ from $x = 0$ to $x = \pi$.

4. Find the surface area of the arc $y = 4x + 1$, $1 \leq x \leq 3$, rotated about the x-axis.

5. Find the surface area of the arc $y = 4x + 1$, $1 \leq x \leq 3$, rotated about the y-axis.

6. Find the surface area of the arc $y = x^2$, $1 \leq x \leq 4$, rotated about the x-axis.

7. Find the surface area of the arc $y = x^2$, $1 \leq x \leq 4$, rotated about the axis $y = -3$.

8. Find the surface area of the arc $y = x^2$, $1 \leq x \leq 4$, rotated about the y-axis.

9. Find the surface area of the arc $y = x^2$, $1 \leq x \leq 4$, rotated about the axis $x = -5$.

10. Find the surface area of the arc $y = x^2$, $1 \leq x \leq 4$, rotated about the axis $x = 5$.

11. Compute the surface area of the surface generated by rotating the arc $y = x^2$ between $x = 1$ to $x = 2$ about the line $x = -4$.

12. **(Gabriel's Horn)** Let $y = \frac{1}{x}$ on the interval $[1, \infty)$. Let R be the region between the graph and the x-axis about the x-axis. Rotate R about the x-axis.

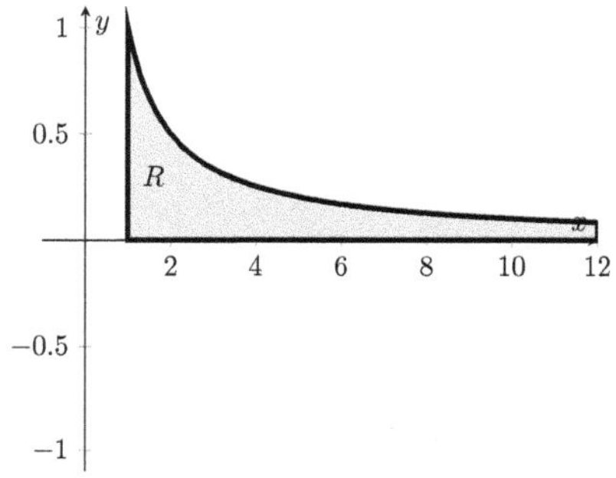

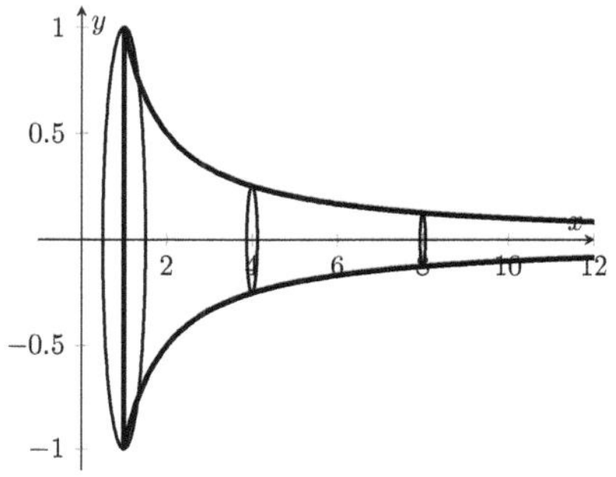

$A =$ The area of R
$V =$ The volume of the generated solid
$S =$ The surface area of the generated surface
Find A, V and S.

Section 20

11.1: Sequences

- Terms about Sequences
  - Increasing sequence
  - Decreasing Sequence
  - Monotonic Sequence
  - Bounded Above Sequence
  - Bounded Below Sequence
  - Bounded Sequence
- Limit Properties
  - $\lim\limits_{n\to\infty}(a_n \pm b_n) = \lim\limits_{n\to\infty} a_n \pm \lim\limits_{n\to\infty} b_n$
  - $\lim\limits_{n\to\infty} c\cdot a_n = c\cdot \lim\limits_{n\to\infty} a_n$
  - $\lim_{n\to\infty} a_n \cdot b_n = \lim_{n\to\infty} a_n \cdot \lim_{n\to\infty} b_n$
  - $\lim\limits_{n\to\infty} c = c$
  - $\lim\limits_{n\to\infty} \frac{a_n}{b_n} = \frac{\lim\limits_{n\to\infty} a_n}{\lim\limits_{n\to\infty} b_n}$
  - $\lim\limits_{n\to\infty} a_n^p = \left(\lim\limits_{n\to\infty} a_n\right)^p$
  - If $\lim a_n = \lim c_n = L$ and $a_n < b_n < c_n$ then $\lim b_n = L$
  - If $f(n) = a_n$ and $\lim\limits_{x\to\infty} f(x) = L$ then $\lim a_n = L$
- Factorials. For example, $6! = 6\cdot 5\cdot 4\cdot 3\cdot 2\cdot 1 = 720$.
- Monotonic Sequence Theorem: Every bounded monotonic sequence converges.

---

1. Determine the next term in the sequence.

   $1, 2, 3, \ldots$

2. Find the next term in the sequence

   $1, 1, 2, 3, 5, 8, \ldots$

3. Find the next term in the sequence

   6517, 6599, 7279, 7844, 7690, ...

4. For each of the sequences, write the first few terms of the sequence and then find $\lim_{n\to\infty} a_n$

   (a) $a_n = 2 + (0.1)^n$

   (b) $a_n = \dfrac{n + (-1)^n}{n}$

   (c) $a_n = \dfrac{1 - 2n}{1 + 2n}$

   (d) $a_n = \sin n$

   (e) $a_n = \sin(1/n)$

   (f) $a_n = \cos(1/n)$

5. Write the first few terms of the sequence and then find $\lim_{n\to\infty} a_n$

   (a) $a_n = \dfrac{\ln n}{n}$

   (b) $a_n = \sqrt[n]{n}$

   (c) Let $x > 0$ be fixed. Let $a_n = x^{1/n}$.

6. Here we find an upper bound for a sequence.
   Let $\{a_n\}_{n=1}^{\infty}$ be the sequence such that $a_n = \frac{3}{n+5}$. Find a number M such that $a_n \leq M$ for all n.

7. For each of the following sequences, determine if it is bounded (above/below/both), monotonic/increasing/decreasing.
It is generally helpful to write out several terms of the sequence to see what is going on.

(a) $a_n = \frac{1}{n}$

(b) $a_n = \frac{-1}{n}$

(c) $a_n = \frac{(-1)^n}{n}$

(d) $a_n = \frac{n+1}{n}$

(e) $a_n = \frac{n+1}{n+2}$

(f) $a_n = n^2$

(g) $a_n = -n^2$

(h) $a_n = (-1)^n n^2$

(i) $a_n = \frac{5n-2}{7n+3}$.

8. Experiment with the following recursive sequences. In particular, write down a few tersm, determine (if possible): monotonicity (increasing/decreasing), bounded (above/below), convergence (converge/ diverge) and find an explicit formula for the terms.

   (a) $a_1 = 4$, $a_n = a_{n-1} + 5$

   (b) $a_1 = 16$, $a_n = \frac{1}{2}a_{n-1}$

   (c) $a_1 = 1$, $a_n = \frac{1}{2}(a_{n-1} + 6)$

   (d) $a_1 = 0$, $a_2 = 1$, $a_{n+1} = a_n + 2a_{n-1}$

   (e) Let $a_1 = 0$, $a_2 = 1$, $a_{n+1} = a_n + 2a_{n-1}$. Now let $b_n = \frac{a_{n+1}}{a_n}$. Find $\lim_{n\to\infty} b_n$.

9. Write the first few terms of the sequence and then find $\lim_{n\to\infty} a_n$ where $a_n = x^n$.

10. Let x be fixed and let $a_n = \left(1 + \frac{x}{n}\right)^n$. Find $\lim_{n\to\infty} a_n$.

11. Let x be fixed and let $a_n = \dfrac{x^n}{n!}$. Find $\lim_{n\to\infty} a_n$.

12. Find $\lim_{n\to\infty} a_n$ where $a_n = \dfrac{n!}{n^n}$.

13. Continued Fractions. Fun and mathematically important, but less important for calculus class.

(a) Find the limit of the sequence:

$$a_1 = 1, \quad a_2 = 1+\frac{1}{1}, \quad a_3 = 1+\cfrac{1}{1+\cfrac{1}{1}}, \quad a_4 = 1+\cfrac{1}{1+\cfrac{1}{1+\cfrac{1}{1}}}, \quad a_5 = 1+\cfrac{1}{1+\cfrac{1}{1+\cfrac{1}{1+\cfrac{1}{1}}}}, \quad \ldots$$

(b) Find the limit of the sequence:

$$a_1 = 2, \quad a_2 = 2+\frac{1}{1}, \quad a_3 = 2+\cfrac{1}{1+\cfrac{1}{2}}, \quad a_4 = 2+\cfrac{1}{1+\cfrac{1}{2+\cfrac{1}{1}}}, \quad a_5 = 2+\cfrac{1}{1+\cfrac{1}{2+\cfrac{1}{1+\cfrac{1}{1}}}},$$

$$a_6 = 2+\cfrac{1}{1+\cfrac{1}{2+\cfrac{1}{1+\cfrac{1}{1+\cfrac{1}{1}}}}}, \quad a_7 = 2+\cfrac{1}{1+\cfrac{1}{2+\cfrac{1}{1+\cfrac{1}{1+\cfrac{1}{4+\cfrac{1}{1}}}}}},$$

$$a_8 = 2+\cfrac{1}{1+\cfrac{1}{2+\cfrac{1}{1+\cfrac{1}{1+\cfrac{1}{4+\cfrac{1}{1+\cfrac{1}{1}}}}}}}, \quad \ldots$$

In this sequence, the "diagonal" of the fractions follows the pattern:
$[2; 1, 2, 1, 1, 4, 1, 1, 6, 1, 1, 8, 1, 1, 10, 1, 1, 12, 1, 1, \ldots]$

Section 21

11.2: Series

- An infinite sum is the limits of the partial sums:

$$\sum_{n=1}^{\infty} a_n = \lim_{n\to\infty} s_n = \lim_{n\to\infty}\left(\sum_{i=1}^{n} a_i\right)$$

- Geometric Series I: $s_N = \sum_{n=0}^{N} ar^n = \frac{a(1-r^N)}{1-r}$

- Geometric Series II: $\sum_{n=0}^{\infty} ar^n = \frac{a}{1-r}$

- Telescoping Series: Use partial fractions (usually) and find an expression for s_N

- Harmonic Series: $\sum_{n=1}^{\infty} \frac{1}{n} = \infty$

- Test for Divergence:
  If $\lim a_n \neq 0$ then $\sum_{n=0}^{\infty} a_n$ diverges
  (No conclusions if $\lim a_n = 0$)

---

1. Find the limit of the following sequence:

$$s_1 = \frac{1}{2}, \qquad s_2 = \frac{1}{2} + \frac{1}{4}, \qquad s_3 = \frac{1}{2} + \frac{1}{4} + \frac{1}{8}, \qquad s_4 = \frac{1}{2} + \frac{1}{4} + \frac{1}{8} + \frac{1}{16}, \qquad \ldots$$

The sequence is given by

$$s_n = \sum_{i=1}^{n} \left(\frac{1}{2}\right)^n$$

2. For a fixed r (real number) and fixed n (positive integer), multiply out and simplify:

$$(1-r)(1+r+r^2+\cdots+r^{n-1}) =$$

3. For a fixed r (real number) and fixed n (positive integer), use long division to simplify

$$\frac{1-r^n}{1-r} =$$

4. Find $\sum_{n=1}^{\infty} \left(\frac{2}{3}\right)^n$

5. Use Geometric series to compute the infinite sums:

(a) $\sum_{n=2}^{\infty} \left(\frac{2}{3}\right)^n$

(b) $\sum_{n=0}^{\infty} \left(\frac{3}{2}\right)^n$

(c) $\sum_{n=2}^{\infty} \left(\frac{2}{3}\right)^{n-2}$

(d) $\sum_{n=0}^{\infty} \left(-\frac{2}{3}\right)^n$

(e) $\sum_{n=2}^{\infty} \frac{2^n \cdot 3^{n-3}}{5^{2n+3}}$

6. Use Geometric series to compute the partial sums

(a) $\sum_{n=1}^{4} \left(\frac{2}{3}\right)^n$

(b) $\sum_{n=1}^{4} \left(\frac{2}{3}\right)^{n-1}$

(c) $\sum_{n=4}^{7} \left(\frac{2}{3}\right)^n$

7. For a given x, such that $|x| < 1$, find $\sum_{n=0}^{\infty} x^n$

8. Consider the repeating fraction

$$\begin{aligned} x &= 3.2414141414141\ldots = 3.2\overline{41} \\ &= 3.2 + \frac{41}{1000} + \frac{41}{100000} + \frac{41}{10000000} + \cdots \\ &= 3.2 + \frac{41}{10^3} + \frac{41}{10^5} + \frac{41}{10^7} + \cdots \end{aligned}$$

Find this sum (as a fraction) using geometric series.

9. Converge or Diverge?

(a) $\sum_{n=1}^{\infty} 10^{500} \left(\frac{99998}{99999}\right)^n$

(b) $\sum_{n=1}^{\infty} \left(\frac{3^n + 7^{n+1} - 23^{n+4}}{5^{2n+1}}\right)$

(c) $\sum_{n=1}^{\infty} \sqrt[n]{2}$

10. Consider the series $\sum_{n=1}^{\infty} \frac{1}{n(n+1)}$.

(a) Find s_9, the sum of the first nine terms.

(b) Find the sum of the series.

11. Use telescoping series to compute the series. Be sure to actually compute a formula for the partial sums, s_N.
I have computed the partial fractions for you.

(a) $\sum_{n=2}^{\infty} \frac{1}{n^2 - n} = \sum_{n=1}^{\infty} \left(\frac{1}{n-1} - \frac{1}{n} \right)$

(b) $\sum_{n=1}^{\infty} \frac{1}{n^2 + 4n + 3} = \sum_{n=1}^{\infty} \frac{1}{2} \left(\frac{1}{n+1} - \frac{1}{n+3} \right)$

(c) $\sum_{n=1}^{\infty} \ln \left(\frac{k+1}{k} \right) = \sum_{n=1}^{\infty} (\ln(k+1) - \ln k)$

Section 22

11.3: Integral Test

Integral Test: $f(x) > 0$, decreasing and $f(n) = a_n$.

1. If $\int_1^\infty f(x)\,dx$ converges then $\sum_{n=1}^{\infty} a_n$ converges.
2. If $\int_1^\infty f(x)\,dx$ diverges then $\sum_{n=1}^{\infty} a_n$ diverges.

---

1. Here we investigate $\sum_{n=1}^{\infty} \frac{1}{n^2}$. Here are some partial sums (computed using WolframAlpha).

$$s_{10} = \sum_{n=1}^{10} = 1 + \frac{1}{2^2} + \frac{1}{3^2} + \cdots + \frac{1}{10^2} \approx 1.5497677311\ldots$$

$$s_{50} = \sum_{n=1}^{50} = 1 + \frac{1}{2^2} + \frac{1}{3^2} + \cdots + \frac{1}{50^2} \approx 1.6251327336\ldots$$

$$s_{100} = \sum_{n=1}^{100} = 1 + \frac{1}{2^2} + \frac{1}{3^2} + \cdots + \frac{1}{100^2} \approx 1.63498390\ldots$$

Based on this information, do you think $\sum_{n=1}^{\infty} \frac{1}{n^2}$ converges or diverges?

2. Here we investigate $\sum_{n=1}^{\infty} \frac{1}{n}$. Here are some partial sums, computed using Wolfram Alpha.

$$s_{100} = \sum_{n=1}^{100} = 1 + \frac{1}{2} + \frac{1}{3} + \cdots + \frac{1}{100} \approx 5.187$$

$$s_{200} = \sum_{n=1}^{100} = 1 + \frac{1}{2} + \frac{1}{3} + \cdots + \frac{1}{200} \approx 5.878$$

$$s_{300} = \sum_{n=1}^{100} = 1 + \frac{1}{2} + \frac{1}{3} + \cdots + \frac{1}{300} \approx 6.283$$

Based on this, does $\sum_{n=1}^{\infty} \frac{1}{n}$ converges or diverge?

3. Converge or Diverge?

   (a) $\sum_{n=1}^{\infty} \frac{1}{n}$

   (b) $\sum_{n=1000}^{\infty} \frac{1}{n}$

   (c) $\sum_{n=1}^{\infty} \frac{1}{n^2}$

   (d) $\sum_{n=965}^{\infty} \frac{1}{n^2}$

   (e) $\sum_{n=0}^{\infty} \frac{2^n}{(n!)^2}$

4. For what values of p does the series $\sum_{n=1}^{\infty} \frac{1}{n^p}$ converge?

5. Determine if the series converge or diverge using the integral test:

   (a) $\sum_{n=0}^{\infty} \frac{1}{1+n^2}$

   (b) $\sum_{n=1}^{\infty} e^{-n}$

   (c) $\sum_{n=2}^{\infty} \frac{1}{n \ln n}$

Section 23

11.4: Comparison Tests

- **Comparison Test For Convergence:**

  If $|a_n| \le |b_n|$ and $\sum |b_n|$ converges, then $\sum |a_n|$ converges.

- **Comparison Test For Divergence:**

  If $|b_n| \le |a_n|$ and $\sum |b_n|$ diverges, then $\sum |a_n|$ diverges.

- **Limit Comparison Test:** Let $L = \lim_{n\to\infty} \frac{a_n}{b_n}$.
  - If $0 < L < \infty$ then:

    $\sum a_n$ and $\sum b_n$ both converge or both diverge.

  - If $L = 0$ and $\sum b_n$ converges, then $\sum a_n$.
  - If $L = \infty$ and $\sum b_n$ diverges, then $\sum a_n$ diverges.

---

1. When n is large, which is bigger:
   $n^2 + 2n + 1$ or $3n^2$?

2. Converge or Diverge? (use comparison test)
   $\sum_{n=1}^{\infty} \frac{n^2 - 2n - 1}{n^4 + 2n + 1}$ compared to $\sum_{n=1}^{\infty} \frac{1}{n^2}$

3. Converge or Diverge? (use comparison test)
   $\sum_{n=1}^{\infty} \frac{n^2 + 2n + 1}{n^4 - 2n - 1}$ compared to $\sum_{n=1}^{\infty} \frac{1}{n^2}$

4. Let $L = \lim_{n\to\infty} \frac{a_n}{b_n}$.
If $L = 1$, which of the following do you expect for large values of n and why?
(a) $a_n < b_n$ (b) $a_n \approx b_n$ (c) $a_n > b_n$

5. Let $L = \lim_{n\to\infty} \frac{a_n}{b_n}$.
If $L = 0$, which of the following do you expect for large values of n and why?
(a) $a_n < b_n$ (b) $a_n = b_n$ (c) $a_n > b_n$

6. Let $L = \lim_{n\to\infty} \frac{a_n}{b_n}$.
If $L = \infty$, which of the following do you expect for large values of n and why?
(a) $a_n < b_n$ (b) $a_n = b_n$ (c) $a_n > b_n$

7. Converge or Diverge? (do a limit comparison test)
$\sum_{n=1}^{\infty} \frac{n^2 - 2n - 1}{n^4 + 2n + 1}$ compared to $\sum_{n=1}^{\infty} \frac{1}{n^2}$

8. Converge or Diverge? (do a limit comparison test)
$\sum_{n=1}^{\infty} \frac{n^2+2n+1}{n^4+2n-1}$ compared to $\sum_{n=1}^{\infty} \frac{1}{n^2}$

9. Determine convergence of the following. Use a comparison or limit comparison (try to use both!).

(a) $\sum_{n=1}^{\infty} \frac{n^2+1}{n^3+1}$

(b) $\sum_{n=1}^{\infty} \frac{2n+1}{(n+1)^2}$

(c) $\sum_{n=1}^{\infty} \frac{1}{2^n-1}$

(d) $\sum_{n=1}^{\infty} \frac{\ln n}{n^{3/2}}$

Section 24

11.5: Alternating Series, Absolute Convergence

- **Alternating Series Test:** Let $S = \sum(-1)^n b_n$ where $b_n > 0$ be an alternating series. S converges if the following are true:
  - $\lim b_n = 0$
  - $b_{n+1} < b_n$ (so the b_n are decreasing)

  And $|R_n| = |S - S_n| \leq b_{n+1}$
- **Absolute Convergence**
  - Definition: $\sum a_n$ is **Absolutely Convergent** if $\sum |a_n|$ converges.
  - Definition: $\sum a_n$ is **Conditionally Convergent** if $\sum a_n$ converges but $\sum |a_n|$ diverges.
  - **Theorem:** If $\sum a_n$ is absolutely convergent then $\sum a_n$ is convergent.
- **Rearrangements:** Don't rearrange conditionally convergent series.

---

1. **Alternating Series:** Does the series converge or diverge:
$$\sum_{n=1}^{\infty} \frac{(-1)^{n+1}}{n} = 1 - \frac{1}{2} + \frac{1}{3} - \frac{1}{4} + \frac{1}{5} + \cdots$$

2. Converge or Diverge? $\displaystyle\sum_{n=0}^{\infty} \frac{(-1)^n}{n!}$
If it converges, find the sum accurate to 0.01.

3. Let $S = \sum_{n=1}^{\infty} \frac{(-1)^{n+1}}{n^2}$.
By the alternating series test, this series converges.

   (a) $S_{15} = 0.8245417573\ldots$.
   How close does the alternating series remainder theorem guarantee S_{15} to be from S?

   (b) If you want the partial sum to be within 0.0001 of the sum, how many terms do you need to take?

4. Let $S = \sum_{n=1}^{\infty} \frac{(-1)^{n+1}}{n^3}$.
By the alternating series test, this series converges.

   (a) $S_{15} = 0.901676118\ldots$.
   How close does the alternating series remainder theorem guarantee S_{15} to be from S?

   (b) If you want the partial sum to be within 0.001 of the sum, how many terms do you need to take?

5. Converge or Diverge? $\frac{1}{4} - \frac{2}{5} + \frac{4}{6} - \frac{8}{7} + \cdots$
   If it converges, find the sum accurate to 0.01.

6. Converge or Diverge? $\sum_{n=1}^{\infty} \frac{(-1)^{n+1}}{\sqrt{n}} = 1 - \frac{1}{\sqrt{2}} + \frac{1}{\sqrt{3}} - \cdots$
   If it converges, find the sum accurate to 0.1.

7. Does the series converge absolutely: $\sum_{i=1}^{\infty} \frac{(-1)^{n+1}}{n^2 + 2n + 1}$

8. Are the following absolutely convergent? Are they conditionally convergent?

(a) $\displaystyle\sum_{n=1}^{\infty} \frac{(-1)^n n^2}{n^2+5}$

(b) $\displaystyle\sum_{n=0}^{\infty} \frac{(-1)^{n-3}\sqrt{n}}{n+4}$

(c) $\displaystyle\sum_{n=2}^{\infty} \frac{\cos(\pi n)}{\sqrt{n}}$

(d) $\displaystyle\sum_{n=1}^{\infty} (-1)^n \sin(\pi/n)$

(e) $\displaystyle\sum_{n=1}^{\infty} \frac{(-1)^n \ln n}{n}$

Section 25

11.6: Ratio and Root Tests

- Ratio Test: Let $L = \lim_{n\to\infty} |a_{n+1}/a_n|$
  - If $L < 1$ then $\sum a_n$ converges absolutely
  - If $L > 1$ then $\sum a_n$ diverges
  - If $L = 1$ then Ratio Test is inconclusive
- Root Test: Let $L = \lim_{n\to\infty} \sqrt[n]{|a_n|}$
  - If $L < 1$ then $\sum a_n$ converges absolutely
  - If $L > 1$ then $\sum a_n$ diverges
  - If $L = 1$ then Root Test is inconclusive

---

1. The series $\displaystyle\sum_{n=0}^{\infty} \left(\frac{2}{3}\right)^n$ converges (why?). What about the following:

   (a) $\displaystyle\sum_{n=0}^{\infty} n \left(\frac{2}{3}\right)^n$

   (b) $\displaystyle\sum_{n=0}^{\infty} n^2 \left(\frac{2}{3}\right)^n$

   (c) $\displaystyle\sum_{n=0}^{\infty} n^3 \left(\frac{2}{3}\right)^n$

   (d) $\displaystyle\sum_{n=0}^{\infty} n^{10000} \left(\frac{2}{3}\right)^n$

2. Let $\sum_{n=0}^{\infty} ar^n$ be a geometric series (so $a_n = ar^n$).

   (a) Compute: $\displaystyle\lim_{n\to\infty} \left|\frac{a_{n+1}}{a_n}\right|$

   (b) Compute: $\displaystyle\lim_{n\to\infty} \sqrt[n]{|a_n|}$

3. Use the Ratio Test to determine convergence.

(a) $\displaystyle\sum_{n=0}^{\infty} \frac{2^n + 5}{3^n}$

(b) $\displaystyle\sum_{n=0}^{\infty} \frac{1}{n^2}$

(c) $\displaystyle\sum_{n=0}^{\infty} \frac{n^3}{(n+1)!}$

(d) $\displaystyle\sum_{n=1}^{\infty} \frac{(2n)!}{(n!)^2}$

(e) $\displaystyle\sum_{n=1}^{\infty} \frac{4^n (n!)^2}{(2n)!}$

4. Use the Root Test to determine convergence.

(a) $\sum_{n=1}^{\infty} \left(\frac{5n - 3n^3}{7n^3 + 2} \right)^n$

(b) $\sum_{n=1}^{\infty} \frac{1}{n^3}$

(c) $\sum_{n=1}^{\infty} \left(\frac{3}{n+1} \right)^n$

(d) $\sum_{n=1}^{\infty} (-1)^{n-1} \left(\frac{2n}{n+1} \right)^n$

(e) $\sum_{n=1}^{\infty} \left(\frac{\ln n}{n} \right)^n$

5. Converge or diverge?

(a) $\sum_{n=1}^{\infty} \frac{(-10)^n}{4^{2n+1}(n+1)}$

(b) $\sum_{n=1}^{\infty} \left(\frac{5n - 3n^3}{7n^3 + 2} \right)^n$

(c) $\sum_{n=0}^{\infty} \frac{n!}{5^n}$

(d) $\sum_{n=1}^{\infty} \frac{n^n}{3^{2n+1}}$

(e) $\sum_{n=0}^{\infty} \frac{(-1)^n}{n+1}$

Section 26

11.3: Series Strategies

Strategies for Testing Convergence/Divergence of a Series:

- Geometric Series
- Telescoping Series
- Test for Divergence
- Harmonic Series
- Integral Test
- p-Series
- Comparison Test
- Limit Comparison Test
- Alternating Series
- Absolute Convergence
- Ratio Test
- Root Test

---

1. What is the difference between a **sequence** and a **series**?

2. What does it mean for a sequence to **diverge**? **Converge?**

3. What does it mean for a series to **diverge**? **Converge?**

4. What is a **partial sum** and what are they good for?

5. Give an example of a convergent **geometric series**.

6. Give an example of a divergent **geometric series**.

7. What is the **Test for Divergence** and what is it good for?

8. What is the **Integral Test** and what is it good for?

9. What is a **p Series**?

10. What is the **Comparison Test**?

11. What is the **Limit Comparison Test**?

12. What is an **Alternating Series**?

13. What is the **Alternating Series Test**?

14. What is **Conditionally Convergent**?

15. What is the **Absolutely Convergent**?

16. What is the **Ratio Test**?

17. What is the **Root Test**?

18. Converge or diverge?

(a) $\sum_{n=2}^{\infty} \frac{n-1}{3n+1}$

(b) $\sum_{n=1}^{\infty} ne^{-n^2}$

(c) $\sum_{n=1}^{\infty} \left(2^{\frac{1}{n}} - 1\right)^n$

(d) $\sum_{n=3}^{\infty} \frac{e^{\frac{1}{n}}}{n^2 - 3}$

(e) $\sum_{n=2}^{\infty} \frac{1}{\ln n}$

(f) $\sum_{n=1}^{\infty} ne^{-n}$

(g) $\sum_{n=2}^{\infty} \left(\frac{n}{n+1}\right)^{n^2}$

(h) $\sum_{n=1}^{\infty} \frac{\pi^{3n}}{n!}$

(i) $\sum_{n=2}^{\infty} \frac{\sqrt{n+1} - \sqrt{n-1}}{n}$

19. Converge or diverge?

(a) $\displaystyle\sum_{n=1}^{\infty} \frac{e^n + n}{e^{3n} - 3}$

(b) $\sum (0.98)^n$

(c) $\sum (-5)^{-n}$

(d) $\displaystyle\sum \frac{1}{n(n+6)}$

(e) $\displaystyle\sum \frac{(-1)^n}{37n}$

(f) $\sum ne^{-n^2}$

(g) $\displaystyle\sum \frac{2^n}{n!}$

(h) $\displaystyle\sum \frac{n!}{e^n}$

(i) $\displaystyle\sum \frac{n^n}{3^{3n+1}}$

20. Converge conditionally, absolutely or diverge?

(a) $\sum_{n=2}^{\infty} \frac{(-1)^n n!}{n^n}$

(b) $\sum_{n=2}^{\infty} \frac{(-1)^n}{n(\ln n)^3}$

21. Which sequences converge?

(a) $a_n = \frac{n^3 + \sqrt{3n^6 + 2n - 1}}{4n^3 + 2n + 5}$

(b) $b_n = \frac{\cos\left(\frac{1}{n}\right)}{n+1}$

22. What does the ratio test tell you about these series?

(a) $\sum_{n=2}^{\infty} \frac{n}{2^n}$

(b) $\sum_{n=1}^{\infty} \frac{\sqrt{n}}{1+n^2}$

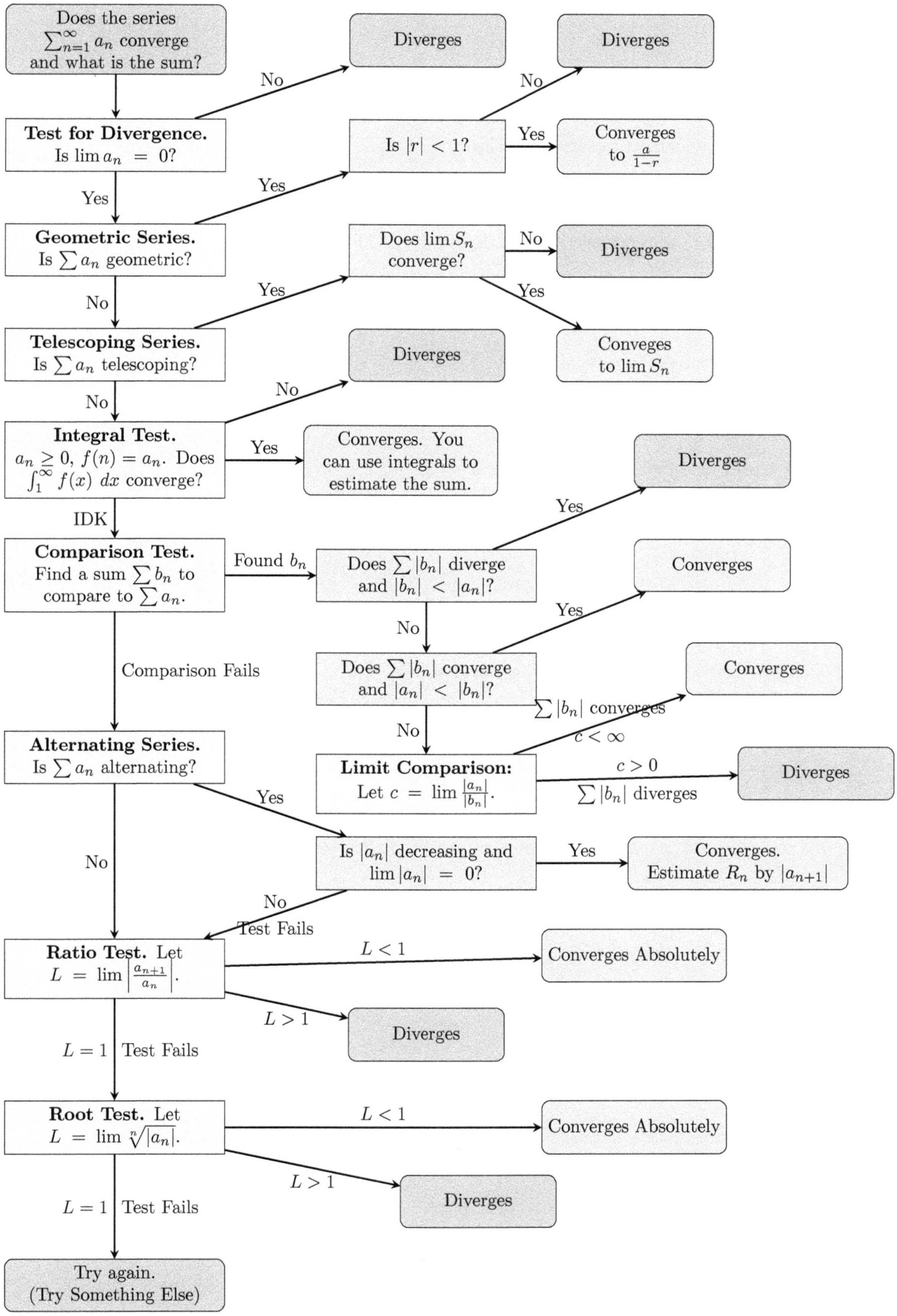

Figure 1: Series Strategy Flow Chart

Section 27

11.8: Power Series

- Goal: Apply what we've learned to series like these:

$$\sum_{n=0}^{\infty} a_n x^n \qquad \text{and} \qquad \sum_{n=0}^{\infty} a_n (x-a)^n$$

- For what values of x does a power series converge?
  - Radius of Convergence
  - Interval of Convergence
- For $\sum a_n (x-a)^n$, there are three possibilities.
  (i) The series converges for only $x = a$.
  (ii) The series converges for all x.
  (iii) There is a positive number R and series converges for $|x-a| < R$ but diverges for $|x-a| > R$.

---

1. Suppose you are to plug different values of x into the power series and compute the sum of the series:

$$\sum_{n=0}^{\infty} x^n$$

Find the easiest x value to plug in to the power series.

2. Suppose you are to plug different values of x into the power series and compute the sum of the series:

$$\sum_{n=0}^{\infty} \frac{1}{n^2+1} (x+1)^n$$

Find the easiest x value to plug in to the power series.

3. Determine the Radius of Convergence and Interval of Convergence for the power series below.

(a) $\sum_{n=0}^{\infty} x^n$

(b) $\sum_{n=0}^{\infty} \frac{(-1)^n x^n}{4^n \ln n}$

(c) $\sum_{n=0}^{\infty} \frac{(-1)^n x^{2n}}{(2n)!}$

(d) $\sum_{n=1}^{\infty} \frac{(-1)^{n+1}(x-1)^n}{n}$

(e) $\sum_{n=0}^{\infty} 2^{2n} x^{2n}$

(f) $\sum_{n=1}^{\infty} \frac{4^n (x-3)^n}{n}$.

Section 28

11.9: Functions as Power Series

- Manupulate known power series to obtain power series for other functions.
  Manipulations:
  - Substitution
  - Addition
  - Multiplication (difficult, used rarely)
  - Division (difficult, used rarely)
  - Differentiation
  - Integration
- Geometric Series. This is the series we will manipulate.

$$\frac{1}{1-x} = \sum_{n=0}^{\infty} x^n = 1 + x + x^2 + x^3 + x^4 + x^5 + x^6 + x^7 + x^8 + \cdots$$

- Some series derived from geometric series:

$$\ln(1+x) = \sum_{n=1}^{\infty} \frac{(-1)^{n+1}}{n} x^n$$

$$\arctan(x) = \sum_{n=0}^{\infty} \frac{(-1)^n}{2n+1} x^{2n+1}$$

- Important note: I *strongly* recommend that you focus on being able to write out the first 3-5 terms of a series rather than trying to find the correct summation formula.

---

1. Given the sum of the geometric series below:

$$\frac{1}{1-x} = 1 + x + x^2 + x^3 + x^4 + x^5 + x^6 + x^7 + x^8 + \cdots$$

What is the sum:

$$1 - x + x^2 - x^3 + x^4 - x^5 + x^6 - x^7 + x^8 + \cdots$$

2. (Method: Substitution)
   Find a series for $\dfrac{1}{1+x^2}$

3. (Method: Substitution)
   Find a series for $\dfrac{1}{4+x^2}$

4. Find a power series for $f(x) = \dfrac{x^3}{4+x}$

5. Let $f(x) = \sum_{n=0}^{\infty} \frac{(-1)^n}{4^{n+1}} x^{n+3}$
Reindex this power series so that the power of x is x^n.
Note: this is something that can be very confusing. First, the most important skill is to be able to write down the first 3-5 terms of a power series correctly. If you insist on writing down the summation formula, then it is usually best to write down the first 3-5 terms and then try to find a pattern.

6. (Method: Differentiation)
Find a series for $\frac{1}{(1-x)^3}$

7. (Method: multiplication)

Use multiplication of power series to find a series for $f(x) = \frac{1}{1-x} \cdot \frac{1}{1-x}$.
This is definitely a pain, but just distribute the terms out and don't worry about finding a pattern–just find the first 4-6 terms of the power series.

$$(1 + x + x^2 + x^3 + x^4 + \cdots)(1 + x + x^2 + x^3 + x^4 + \cdots) =$$

8. (Method: multiplication)
Given the power series for $\ln(1+x)$ and $\frac{1}{2+x}$ below, find a power series for $\dfrac{1}{2+x}\cdot\ln(1+x)$.

$$\ln(1+x) = x - \frac{x^2}{2} + \frac{x^3}{3} - \frac{x^4}{4} + \frac{x^5}{5} - \cdots$$

$$\frac{1}{2+x} = \frac{1}{2} - \frac{x}{4} + \frac{x^2}{8} - \frac{x^3}{16} + \frac{x^4}{32} - \cdots$$

$$\frac{1}{2+x}\cdot\ln(1+x) = \left(x - \frac{x^2}{2} + \frac{x^3}{3} - \frac{x^4}{4} + \frac{x^5}{5} - \cdots\right)\left(\frac{1}{2} - \frac{x}{4} + \frac{x^2}{8} - \frac{x^3}{16} + \frac{x^4}{32} - \cdots\right)$$

9. (Method: Long Division. Challenging and worth avoiding whenever possible.)
Use long division of power series to find a series for $\frac{\ln(1+x)}{\ln(1-x)}$
Note: $\ln(1+x) = x - \frac{1}{2}x^2 + \frac{1}{3}x^3 - \frac{1}{4}x^4 + \frac{1}{5}x^5 - \cdots$
and: $\ln(1-x) = -x - \frac{1}{2}x^2 - \frac{1}{3}x^3 - \frac{1}{4}x^4 - \frac{1}{5}x^5 - \cdots$

10. Find power series for the following functions.

(a) $\dfrac{1}{4+3x}$

(b) $\dfrac{x}{(1+x)^2}$

(c) $\dfrac{1+2x}{1-x}$

11. (Method: integration)
Find a power series for $\ln(1+x)$.

12. (Method: Integration)
Find a series for $\arctan x$.
Note $\int \frac{1}{1+x^2}\,dx = \arctan x + C$

13. Use the power series for $\arctan x$ to find a series for $\arctan(1) = \pi/4$.

14. Find a series for $\frac{1}{1+x^{10}}$.
Use your series to approximate $\int_0^{0.25} \frac{1}{1+x^{10}}\,dx$ to within 10^{-7}.

Section 29

11.10: Taylor Series

- Goal: Given a function: find a power series that equals the function.
- Taylor Series.

$$f(x) = \sum_{n=0}^{\infty} c_n (x-a)^n \text{ where } c_n = \frac{f^{(n)}(a)}{n!}$$

- $T_n(x)$ is the Taylor Polynomial of degree n.
- Maclaurin Series: Taylor Series with $a = 0$.

$$f(x) = \sum_{n=0}^{\infty} c_n x^n \text{ where } c_n = \frac{f^{(n)}(0)}{n!}$$

- Vocabulary. The following are often used synonymously (with sentences for context):
  - Series: "Find a series for $\sin x$."
  - Power Series: "Find a power series for $\sin x$."
  - Taylor Series: "Find the Taylor Series for $\sin x$."
  - Mclaurin Series: "Find the Mclaurin Series for $\sin x$."
- Important Series

$$\frac{1}{1-x} = \sum_{n=0}^{\infty} x^n = 1 + x + x^2 + x^3 + x^4 + \cdots$$

$$e^x = \sum_{n=0}^{\infty} \frac{x^n}{n!} = 1 + x + \frac{1}{2!}x^2 + \frac{1}{3!}x^3 + \frac{1}{4!}x^4 + \cdots$$

$$\sin x = \sum_{n=0}^{\infty} (-1)^n \frac{x^{2n+1}}{(2n+1)!} = x - \frac{1}{3!}x^3 + \frac{1}{5!}x^5 - \frac{1}{7!}x^7 + \cdots$$

$$\cos x = \sum_{n=0}^{\infty} (-1)^n \frac{x^{2n}}{(2n)!} = 1 - \frac{1}{2!}x^2 + \frac{1}{4!}x^4 - \frac{1}{6!}x^6 + \cdots$$

$$\ln(1+x) = \sum_{n=1}^{\infty} (-1)^{n-1} \frac{x^n}{n} = x - \frac{x^2}{2} + \frac{x^3}{3} - \frac{x^4}{4} + \cdots$$

$$\arctan(x) = \sum_{n=0}^{\infty} (-1)^n \frac{x^{2n+1}}{2n+1} = x - \frac{x^3}{3} + \frac{x^5}{5} - \frac{x^7}{7} + \cdots$$

$$(1+x)^k = \sum_{n=0}^{\infty} \binom{k}{n} x^n$$

- Taylor Remainder Theorem (if applicable, it is easier to use alternating series remainder):
  If $\left|f^{(n)}(x)\right| \leq M$ for all $|x-a| \leq d$ then

$$|R_n(x)| \leq \frac{M}{(n+1)!}|x-a|^{n+1} \qquad \text{for } |x-a| \leq d$$

1. Let $f(x) = \sum_{n=0}^{\infty} \frac{x^n}{n!}$

   (a) Find $f(0)$

   (b) Find $f(1)$

   (c) Find $f'(x)$

   (d) Find $f''(x)$

   (e) Graph $f(x)$.

2. Find the Taylor series, centered at $x = 0$, for $f(x) = e^x$.
I suggest you do this by completing the table below. (Get used to make such tables! They will be very helpful!)

n	$f^{(n)}(x)$	$f^{(n)}(0)$	$c_n = \frac{f^{(n)}(0)}{n!}$
0	e^x		
1			
2			
3			
4			
5			
6			
n			

3. Find the Taylor series, centered at $x = 0$ for $f(x) = \ln(1 + x)$.

4. Find the Taylor Series for $f(x) = \ln x$ centered at $x = 1$.

5. Find the Taylor Series for $f(x) = \sin x$ centered at $x = 0$.

6. Find the Taylor Series for $f(x) = \cos x$ centered at $x = 0$.

7. Find the Taylor Series for $f(x) = 4 + 2x - 3x^2 - x^3 + 7x^4 - x^5$ centered at $x = 0$.

8. Find the Taylor Series for $f(x) = 4 + 2x - 3x^2 - x^3 + 7x^4 - x^5$ centered at $x = 1$.

9. Let $f(x) = \cos(x^2)$

   (a) Find the Taylor Series for $f(x)$ centered at $x = 0$

   (b) Compute $\int_0^{1/2} \cos(x^2)\, dx$ using the Taylor series.

10. Multiply to find a power series for $(\sin x)(\cos x)$:

11. Use multiplication of power series to find a power series for $e^x \sin x$.

12. Use division of power series to find a power series for $\frac{\sin x}{x}$.

13. Use power series to find $\lim\limits_{n\to\infty} \frac{\sin x}{x}$.

14. Here we experiment with the series for e^x and complex numbers.

    (a) Write down the series for e^x. Write down at least 6-7 terms.

    (b) Using substitution, write down the series for e^{ix}. Write down at least 5-6 terms. You will need to simplify this.
    Remember that $i = \sqrt{-1}$, so $i^2 = -1$, $i^3 = -1$, $i^4 = 1$, $i^5 = -1$, etc.

15. Let $f(x) = \dfrac{e^x - 1 - x}{x^2}$.

(a) Write down the Mclaurin series for $f(x)$.

(b) Using your power series, find $\lim\limits_{x \to 0} \dfrac{e^x - 1 - x}{x^2}$.

16. Find the Taylor series for $f(x) = 1 + x + x^2$ centered at $x = 2$.

17. Given a function, $f(x)$, it is always possible to find a Taylor Series that equals $f(x)$.
What this is asking is the following: If you start with a function, can you make a table of derivatives, letting $c_n = f^{(n)}(a)/n!$ and get a series that is equal to the function?

18. The degree 2 Taylor polynomial for $\cos x$ centered at 0 is $T_2(x) = 1 - \frac{x^2}{2!}$.
Find an estimate for the maximum error (given by remainder theorem) for $|R_2(0.2)| = |\cos(0.2) - T_2(0.2)|$.

19. Let $f(x) = \sqrt{x}$

    (a) Find the Taylor series centered at 4 for $\sqrt{x}$. Write out at least to the degree 3 terms.

    (b) Use $T_2(x)$ to approximate $\sqrt{4.03}$.

    (c) Estimate the error, $R_2(4.03)$, for $T_2(4.03)$.

www.ingramcontent.com/pod-product-compliance
Ingram Content Group UK Ltd.
Pitfield, Milton Keynes, MK11 3LW, UK
UKHW050722170926
15229UKWH00010B/43

9 781524 978501